Erosion and Sediment Pollution Control

Erosion and Sediment Pollution Control

R. P. BEASLEY

THE IOWA STATE UNIVERSITY PRESS, *Ames, Iowa*

R. P. BEASLEY was the recipient in 1971 of the Hancock Soil and Water Engineering Award presented annually to a member of the American Society of Agricultural Engineers for noteworthy contributions to the advancement of soil and water engineering. The award recognized Professor Beasley's influence as a teacher and his development of improved practices for the control of erosion during the past three decades.

Appointed to the Agricultural Engineering staff at the University of Missouri in 1941, Professor Beasley was advanced to his present rank in 1955. He has served on many regional and national committees and was chairman of the Soil Erosion Group of ASAE. He is the author of numerous technical publications, a registered professional engineer in Missouri, and a member of a number of scientific and honorary societies.

Library of Congress Cataloging in Publication Data
Beasley, Robert Patrick.
    Erosion and sediment pollution control.
    Includes bibliographies.
    1. Soil conservation. 2. Soil erosion. I. Title.
S623.B33   631.4'5   79–153158
ISBN 0–8138–1530–4

Composed and printed by
The Iowa State University Press

First edition, 1972
Second printing, 1973
Third printing, 1974
Fourth printing, 1976
Fifth printing, 1978

CONTENTS

∼ PREFACE

This book is designed as a textbook and a source of information for vocational agriculture teachers, Extension and Soil Conservation Service personnel, and others concerned with agricultural or urban erosion control.

SOIL is one of the vital substances necessary to human existence. Soil erosion depletes the productivity of our soils and produces sediment which is one of the major pollutants of our environment.

Many new concepts and practices have been developed for the control of erosion during recent years. It is important that this information be made available to students and to those engaged in the many aspects of soil and water conservation.

This book gives consideration to the effects of soil erosion on civilization in the past and the problems it creates in the present. It explains how erosion occurs and evaluates the various factors affecting it. Mechanical, agronomic, and management practices that can be used to reduce erosion are explained in detail. The need for integrating these practices into a plan to control erosion, improve the productivity of the land, increase the efficiency of the farm as a production unit, and reduce sediment pollution problems is emphasized.

The primary objective of this book is to help conservationists develop the ability to analyze a given situation, determine the practices necessary for effective erosion control, and put these practices into effect on the land. Sufficient information is given in hydrology, hydraulics, and surveying so that this objective can be accomplished.

This book is based on the author's thirty years of experience in teaching and research and his close association with Extension and Soil Conservation Service personnel and contractors engaged in soil and water conservation work. It is the outgrowth also of two previous texts co-authored with Professor J. C. Wooley: *Farm Water Management* (1950) and *Farm Water Management for Erosion Control* (1957).

The author wishes to acknowledge the assistance of fellow workers in soil and water conservation whose efforts have resulted in the development of many of the methods of control presented in this book. Special acknowledgment is due Neil P. Woodruff, Agricultural Engineer, Agricultural Research Service, USDA, for his assistance in providing information for and reviewing those sections of the book on wind erosion and its control. The author is especially grateful to the secretarial staff of the Agricultural Engineering Department, University of Missouri, Columbia, for their assistance in the preparation of material which was used for class instruction while the manuscript for this book was being developed.

Erosion and Sediment Pollution Control

1 ~ Man and the Land

*In a study of soil erosion and sediment
pollution control, it is profitable to
review the record of man's use of the land.
Such a review reveals successes and
failures, and from these, knowledge may
be gained which can be applied to the
solution of present day problems.*

SOME 350 million years ago, plants and animals previously confined to the sea began to establish themselves on the land. They had difficulty at first in securing a foothold on the barren rocks. This was a period of survival of the fittest; no species of plant or animal could survive if it did not adapt to the new environment. As the plants and animals died, their remains added to the body of the soil, and through millions of years a thin mantle of soil was accumulated (2).

When primitive man came into the picture he, like other animals, was forced to live in a manner that did not upset the balance that nature demanded for survival. As man became civilized, his demands upon the land for additional food and clothing grew. He changed from nomadic herding and began cultivating the soil. This intensified use of the land resulted in the destruction of the natural cover and exposed more of the soil to the erosive forces. In some cases, man was able to control these erosive forces and establish a permanent agriculture. In many other cases however he was not able to control erosion, and devastation of the land resulted. It is probable that since the dawn of history, man has destroyed as many productive acres as now exist in the world (3). This destruction has contributed greatly to the decline and fall of civilizations (5). In a few limited cases, man has been able to control erosive forces and establish an agriculture that has survived through the centuries.

Mesopotamia. The Tigris and Euphrates rivers in Mesopotamia once watered a land so rich as to suggest this as the location of the

3

Garden of Eden. A vast irrigation system built 2,000 years before Christ made possible an agriculture which supported great cities and a complex civilization. But the forests on the hills surrounding the valley were cut down and erosion debris flooded into the rivers and irrigation ditches. Today Mesopotamia is a desert, a drifting waste of sand which drowns the ruins of mighty Babylon and great buildings and monuments.

Syria. Syria was once considered a land of great fertility. Dense forests grew on slopes covered with rich red-brown soil and provided the lumber for ships and cities. The land was cleared and cultivated and for a time more than one hundred cities prospered. In a few centuries, erosion had stripped away the topsoil until today the doorsteps of ruined cities stand several feet above the unproductive substratum. This area, which once exported enormous quantities of wine and olive oil to Rome, is now a desolate wasteland. The story repeated itself in many countries as civilization moved eastward to China and westward through Europe and across the Atlantic Ocean to the Americas.

China. The uplands drained by the Hwang-Ho, or Yellow River, of China were once very rich and the home of a numerous and prosperous people. They are now an eroded desolation, and the light-colored subsoil washing downstream gives the Yellow River its name. During flood stage this eroded soil accounts for 50 percent by weight of the volume of flow of the river (11). Not only has erosion ruined the uplands but it has created tremendous problems downstream where the river flows for 400 miles across a delta. This eroded soil has so built up the river's channel that it stands 40 to 50 feet above the surrounding delta. The river is precariously contained by earthen dikes which have been built by millions of Chinese using bare hands and baskets throughout thousands of years. However, these dikes break every so often, resulting in the drowning of thousands of people and the ruin of many desperately needed acres.

Recently China has attempted to reclaim some of the eroded uplands. A picture in *Time* magazine, December 1, 1958, captioned "Ant army of 150,000 peasants fighting erosion in the Kansu province," shows the 150,000 men attempting to build terraces on a steep mountain slope with shovels and hoes.

Lebanon. In Lebanon the mountains were once covered by the famous forest, "The Cedars of Lebanon." Nearly 3,000 years ago King Solomon made an agreement with Hyram, King of Tyre, to furnish him with lumber for construction of the temple and other structures in Jerusalem. Solomon supplied 80,000 lumberjacks to work in the forest and 70,000 workmen to skid the logs to the sea. As the timber was removed, the land was placed under cultivation, and the soil began forthwith to erode. In most areas little effort was made to control erosion and the country was devastated. In isolated areas efforts were made to control erosion by construction of rock-walled bench terraces across the slope. Farming has been carried on successfully on these slopes for thousands of years.

South America. In the lofty plateaus of the Andes, farmers are still farming the steep mountain slopes that were farmed by their Inca ancestors thousands of years ago. The fields have been carefully terraced to catch rainfall and prevent erosion. These vast works, an arresting monument to the tillers of this soil throughout thousands of years, show the length to which some people will go to save their soil when the necessity for food requires it.

Holland. Another example of the length to which a nation will go to provide food for her people is provided by the Netherlands—in this case, the reclamation of land from the sea. The source of the following is an excellent book, *Dredge, Drain, Reclaim,* by Dr. Johan van Veen, an outstanding engineer and the late Chief Engineer, Rjkswaterstaat (15).

The Netherlands may be said to be a sand and mud dump left over from the ice age. Some early explorers called this seacoast the Gates of Hell; others described it as somewhere in the north, where ice, water, and air mingled, without doubt the end of the earth. Archaeological excavations indicate that about 4000 B.C. a group of farmers had been forced from the interior of the country into this marshland. In order to keep above water, they carried mud by hand in baskets and made artificial mounds on which they built their homes and produced food. In all they made 1,260 of these mounds in the northeastern Netherlands and more in East Friesland. The areas of the mounds vary from 5 to 40 acres. A single mound may contain up to 1 million cubic yards of earth.

Later these people built seawalls and dikes to hold out the sea so that the lower areas could be farmed. Up to 1860, just before the advent of steam dredging, they had built over 1,750 miles of them containing some 250 million cubic yards of earth, practically all transported by hand or in later years by horse drawn carts. The third great work was the digging of ditches and canals from which some 200 million cubic yards of earth were removed. The fourth and greatest task was the digging of peat which provided fuel. When the peat was removed, lakes were created which, when drained, provided additional fertile land to farm.

Netherlanders have dug by hand the enormous volume of some 10 billion cubic yards of earth. This is one hundred times the volume of earth removed by sixty steam dredges in the construction of the Suez Canal. The people of the Netherlands continue to reclaim land from the sea with the most efficient dredging fleet in the world. Since the year 1200 they have reclaimed some 1,586,000 acres, and another 257,-000 acres are in the process of reclamation. Balanced against this however is the loss of some 1,400,000 acres to the sea during this period.

The Netherlands today is a progressive country with nearly nine hundred people per square mile. It has the highest longevity of the world, the lowest death rate, the highest yield of cereals and potatoes per acre, the highest production of milk per cow, and one of the highest national incomes per unit of area. This all from a saltwater swamp which would have been considered by many as a habitat only for wild ducks, gulls, and mosquitos.

Why have some nations with limited resources been able to grow and prosper, while others with a greater abundance of resources have not? The answer involves the attitude, philosophy, customs, and traditions of the people and the social, economic, religious, and political institutions they have developed (4). An ancient motto which is hewn into a stone standing on the Zuider Zee dam may give some indication of why the Netherlands has prospered. This motto, "A Nation That Is Alive Builds for Its Future," indicates a well-organized, intelligent, and conscientious effort on the part of the people to improve their land, not only to meet present needs but also to provide for future generations.

The United States. The American Indian, as is true of most primitive people, had very close ties to his natural habitat and had a deep reverence, pride, and affection for the land. The Indians considered themselves to be a part of nature, not as rulers of nature or as owners of the land.

The philosophy of most white men who came to America was quite different. They coveted land for the wealth it would produce. With little regard for the interrelationships of man and his earth, they considered land as something to own for profit making only.

The Indians and white men were congenial the first few years because each had something to learn and to gain from the other. However, as the white people wanted more and more land, trouble began. The Indians were forced to give up the land when it became more valuable to the white men than their friendship (10).

What has been the history of the white man's use of this land? The early settlers were mostly from western Europe. They were familiar with farming methods in that area where a majority of the rains were of such low intensity that runoff and erosion did not create a problem. In this country they encountered intense rainfall and steep slopes. They adopted new soil-exposing row crops—corn, cotton, and tobacco—and continued to use west European farming methods. There was seemingly an unlimited supply of land which enabled them to devastate a farm and move on with the attitude: "You can't tell me how to farm; I have already worn out two farms." Large machines were developed which enabled the farmers to exploit larger and larger areas of land. The combination of all these factors has resulted in the highest rate of destruction of the largest area of productive soil in the history of man.

The extent of erosion in this country has been determined by a nationwide survey made by the Soil Erosion Service in 1934. At that time there were approximately 50 million acres of land from which topsoil had been eroded, and the land was so riddled with gullies that it was virtually useless for crop production. (For comparison of areas, there are approximately 44 million acres in the state of Missouri.) There were approximately 280 million acres of formerly productive land that had been so severely damaged by soil erosion that their use for cropland or grazing was not economically feasible. A total of approximately 775 million acres had become so severely eroded as to require erosion control measures to ensure continued productivity.

In 1965 the results of a national inventory of soil and water conservation needs were published (12). Following is a statement from this publication: "Susceptibility to soil erosion is the most widespread conservation problem limiting the land capability. It is the dominant problem on about one-half (51 percent), or 738 million acres of the area included in the Conservation Needs Inventory."

A national inventory of soil and water conservation needs was conducted by several federal agencies in 1967. The following information was taken from a preliminary report of this inventory (14): "Fifty-nine percent of the cropland, 66 percent of the native grazing land, 68 percent of the pastureland, and 71 percent of the commercial forest land in the United States needs additional soil and water conservation treatment."

The 1967 inventory included all soil and water conservation needs, but soil erosion was still the dominant problem. For example, the inventory indicated that cultural or mechanical practices need to be applied on 276.8 million acres—more than 60 percent of the cropland—to reduce erosion losses to an acceptable minimum. The results of these surveys indicate that there has been slight progress in soil erosion control on land used for crop production during the past thirty years. However, soil erosion on sloping land is still a major problem which must be solved if we are to establish an enduring agriculture and provide adequate food for the future.

Erosion is not limited to farmlands. Severe erosion has occurred and is occurring on lands used for urban development, highways, roads, and strip-mining. Soil erosion has not only reduced land available for food production, but has produced sediment which is a major source of pollution of our streams and lakes.

One of the basic problems facing the world in the future will be the pressure of an increasing population on land resources. The area of land is fixed but the human race is expanding at a rate that makes population difficult to predict. The land available for food production however is not fixed, since the increasing population makes many other demands on land. Construction of homes, industries, and highways is only one of the many uses that are taking land out of production.

It has been estimated that by the year 2000, the food output must triple that of 1965 to avoid widespread starvation and considerably more than that to provide an adequate diet (1). The increased demand for grain crops will result in the further expansion of crop production on erodible upland soil. It seems imperative that we must prevent further decline of the productivity of these soils by erosion if we are to meet the demands for increased food production.

The United States has been blessed with an abundance of natural resources, and its people have been able to convert these resources into wealth and affluence. Our society however is depleting these resources, and people are becoming concerned with the decrease in open spaces available and the pollution of streams and lakes. We fail to realize that our demands for affluence contribute to the deterioration of our natural resources. We demand large volumes of newsprint and blame the paper mills for polluting the streams. We are so rich in resources that we condone an economy of planned obsolescence that perpetuates a waste of natural resources. We discard clothing that is not worn out, but only out of style. We throw out more food in our garbage than some people have to eat. Conservation cannot come of age until we realize our individual and collective roles in preserving the environment (9).

One of the greatest challenges of our time is to find ways and means of applying our machines, our science, and our technology to the problem of supplying the food needed and providing the quality of environment desired; and at the same time not only conserving but increasing the productivity of our soil resources.

PROBLEMS

1.1.   Why has the history of the use of land in the United States generally been one of soil exploitation rather than one of soil conservation?

1.2.   Has the policy of land exploitation followed by the United States been good or bad for the total national welfare to date?

1.3.   What should our future policy be in regard to soil conservation and how should this policy be implemented?

REFERENCES

1. Burton, Ian, and Kates, R. K. *Readings in resource management and conservation*. Chicago: Univ. Chicago Press, 1965.
2. Dale, T., and Carter, V. G. *Topsoil and civilization*. Norman: Univ. Okla. Press, 1955.
3. Eisenhower, Milton S. Man, soil and civilization. An address given before the Agriculture, Industry, and Science Conference, Hays, Kansas, April 1948.
4. Hockensmith, R. D. Conservation—Ingredient of world peace and progress. *J. Soil Water Conserv.*, vol. 17, no. 6, November–December 1962.
5. Lowdermilk, W. C. Conquest of the land through 7,000 years. USDA Inf. Bull. 99, 1953.
6. Parson, Ruben L. *Conserving American resources*. Englewood Cliffs, N.J.: Prentice-Hall, 1964.
7. Stallings, J. H. *Soil conservation*. Englewood Cliffs, N.J.: Prentice-Hall, 1964.
8. Thompson, Louis M. Impact of world food needs on American agriculture. *J. Soil Water Conserv.*, vol. 23, no. 1, 1968.
9. Towell, W. E. Wise land use. *J. Soil Water Conserv.*, vol. 24, no. 2, March–April 1969.

10. Udall, Stewart L. *The quiet crisis.* New York: Avon, 1963.
11. United Nations. *Methods and problems of flood control in Asia and the Far East.* Bangkok: Bureau of Flood Control of the Economic Commission for Asia, 1951.
12. USDA. Soil and water conservation needs—A national inventory. Misc. Publ. 971, 1965.
13. ———. *Soils and men.* Yearbook of Agriculture, 1938.
14. ———. Soil and water conservation needs inventory. *Soil Conserv.,* vol. 35, no. 5, December 1969.
15. Veen, Johan van. *Dredge, drain, reclaim.* 5th ed. The Hague, Netherlands: Martinus Nijhoff, 1962.

2 ~ Water Erosion

*Erosion by water is the result of energy
developed by the water as it falls
toward the earth and as it flows over the
surface of the land. In the erosion process,
soil particles are detached from the soil
mass and transported to another location.*

FIRST CONSIDER the energy present in an intense rainstorm. It is not unusual to have a rainstorm producing 2 inches of rain in a thirty-minute period. These 2 inches of water on an acre of land would weigh 453,000 pounds or 226.5 tons. This rainstorm would have an intensity of 4 inches per hour. The mean diameter of the raindrops would be approximately 0.12 inch and they would be falling at a velocity of approximately 25 feet per second. The energy created by this amount of water falling at this velocity must be absorbed by the soil.

RAINDROP EROSION

The damage caused by raindrops hitting the soil at a high velocity is the first step in the erosion process. We may think of the raindrops as miniature bombs hitting the soil surface. They shatter the soil granules and clods, reducing them to smaller particles and in turn reducing the infiltration capacity of the soil. A raindrop hitting on a wet soil forms a crater, compacting the area immediately under the center of the drop, moving detached particles outward in a circle around this area, finally meeting sufficient resistance to be deflected upward. Authorities have estimated that more than 100 tons of soil per acre may be detached by the raindrops in a single rain (7). The splashed particles may reach a height of 2 or 3 feet and may cover a radius of as much as 5 feet. On sloping land, the force of the raindrops on the surface is such that more than half of the splashed soil is moved downslope as it falls back to the surface. The force of millions

10

of raindrops during an intense rain on a cultivated field results in an appreciable movement of soil downslope.

Raindrops hitting a soil surface that is covered by a film of water churn up the soil so that the surface film of water becomes quite muddy. As this muddy water infiltrates into the soil, the particles are filtered out in the surface layer. The infiltration of the muddy water and the compacting and puddling action of the raindrops falling on the surface combine to form a layer of soil which has a much lower infiltration rate than that of the soil at the beginning of the rain. This effect is most pronounced for rains of high intensity.

When the rate of rainfall exceeds the rate of infiltration, the depressions on the surface fill and overflow to cause runoff. During the rain the runoff is splashed and resplashed millions of times by the falling raindrops. This breaks the soil particles it is carrying into smaller and smaller sizes and helps to keep them in suspension. Thus a thin sheet of water will carry a much heavier load of silt during a rain than it would if the splashing action were not present.

A good example of this is given by a practical illustration. A parking lot had just been paved but the grading around it had not been completed. A heavy rain eroded the clay, and muddy water ran across the paved lot. About 0.5 inch of clay was deposited under each parked car, but the area between cars exposed to the rain was washed clean.

Falling raindrops break down soil aggregates, detach and transport soil particles, keep the runoff water loaded with the finer, more valuable particles, and cause sealing and compacting of the soil surface. This reduces the ability of the soil to absorb water and increases surface runoff.

SHEET EROSION

Soil movement resulting from raindrop splash and surface runoff is often called sheet erosion. Sheet erosion removes the lighter soil particles, organic matter, and soluble nutrients from the land and is thus a serious detriment to the maintenance of soil fertility and productivity. Since sheet erosion occurs rather uniformly over the slope, it may go unnoticed until most of the productive topsoil has been removed. For this reason, sheet erosion must be considered as most serious.

As the surface water accumulates, it moves downslope. This water rarely moves as a uniform sheet over the surface of the land. It would move in this manner if the surface were smooth and uniformly inclined, which is seldom the case. The surface is almost always irregular. Surface areas a few feet square generally exhibit in miniature the drainage pattern of a major watershed. Each small portion of the runoff water takes the path of least resistance, concentrating in depressions and gaining in velocity as the depth of water and the slope of the land increases.

The erosiveness of flowing water depends upon its velocity, turbulence, and the amount and type of abrasive material it transports. Velocity increases as the depth of flow and the slope of the land increases. Turbulence of flow increases as the rainfall becomes more intense and as the surface flow concentrates in depressions. Abrasive capacity of the runoff depends upon the energy of the flowing water and the amount and type of suspended material in the water.

Soil particles are detached by a combination of rolling, lifting, and abrasive action. When flowing water moves over a soil surface, horizontal forces act upon the particles in the direction of flow. These forces detach particles from the soil mass by rolling or dragging them out of position. As the surface flow concentrates in depressions, the flow becomes more turbulent, and the different velocities and pressures cause vertical currents and eddies. The upward movement of the water past the soil particles detaches them by a lifting action. Soil detachment by abrasion occurs when particles already in transit in the flow strike or drag over particles on the soil surface and set them in motion.

Soil particles are transported by a combination of surface creep, saltation, and suspension. The horizontal forces of water flowing over the surface transport soil particles by rolling or sliding them along in contact with the land surface. This is called surface creep. Movement by saltation occurs when forces due to turbulence lift the particles from the surface and move them along by a continuous series of steps or jumps. When the upward velocities in the flow exceed the settling velocities of the detached particles, transportation by suspension occurs. Particles transported by suspension may travel long distances before settling to the land surface.

The amount of material transported depends upon the transporting capacity of the runoff and the transportability of the soil. The transportability of the soil is influenced by the size, density, and shape of the individual soil particles, and by the retarding effect of vegetation and obstructions. Soil that is moved downslope by the action of the raindrops and by shallow flow over the soil surface consists of the smaller and lighter soil particles. The larger and heavier soil particles are more difficult to transport, hence they are not moved as great a distance.

RILL EROSION

As the surface runoff concentrates in surface depressions sufficient soil may be removed to form small but well-defined channels. If these channels do not seriously interfere with normal tillage operations, they are called rills. Since small channels or rills are obliterated by normal tillage operations, this type of erosion may go unnoticed until serious damage to productivity has resulted.

GULLY EROSION

When surface channels have been eroded to the point that they cannot be smoothed over by normal tillage operations, they are called gullies. Gullies may develop as a result of several factors.

Channels. Soil is removed by surface water concentrating in and flowing through surface channels in sufficient volume to form a gully. Gully formation by this process is usually relatively slow, particularly where the soil is fairly resistant to erosion.

Waterfalls. Water from surface channels is often discharged over an abrupt change in grade. This stream of water falling to a lower elevation has greatly increased eroding power as compared to the same stream on a uniform grade. The channel at the foot of the waterfall is deepened, and the banks are undermined and cave in. Gullies formed by this process may be quite deep; some in the deep loess soils attain depths of over 60 feet.

Freezing and thawing. Alternate freezing and thawing of the exposed gully banks result in sloughing of the sides and enlargement of the gully.

Slides and mass movement of soil. On gully banks gravitational and seepage forces tend to cause movement of the soil mass from a higher to a lower elevation. These forces induce a shearing stress within the soil mass which slips into the gully if the soil does not have sufficient resistance to these forces. The soil which has been detached and transported by gully erosion is of less value than soil removed by sheet erosion because it contains a higher percentage of subsoil. It costs much more to bring this type of erosion under control, but it is much easier to get action in establishing protective measures. However, the farmer and society pay a severe penalty for allowing erosion to develop to this stage.

CHANNEL EROSION

Large channels often carry runoff for an appreciable length of time following a rain. The running stream cuts into the bank below water level, finally undermining it to the extent that it will cave or slide into the stream bed. The tendency of streams to meander increases stream channel erosion and results in some cases in an appreciable loss of land.

DEPOSITION

Deposition is the end of the erosion process. The soil particles come to rest when resistance forces are greater than erosive forces. Soil materials deposited by flowing water are usually separated by particle size. The first to be deposited are the heavier particles of lowest transportability, whereas the lighter particles of highest transportability are deposited farther downstream. The deposition of the erosional debris in places where it is not wanted is one of the major damages resulting from erosion. Fertile soils below upland slopes and in bottomlands are often covered with the less fertile material carried by the runoff. Crops and pastures are often damaged by soil deposited on them by the runoff water. Stream channels are clogged and reservoirs reduced in capacity by the deposition of sediment. Pollution by sediment is one of the major factors causing a deterioration in the quality of our streams and lakes.

PROBLEMS ASSOCIATED WITH WATER EROSION

As a nation becomes more industrialized and urbanized, it is increasingly difficult for its people to realize their dependence on the land and the effect that soil erosion may have on the future of their lives and their nation. Soil erosion affects our lives in so many ways that it is difficult to comprehend the magnitude of the problem.

LOSS OF SOIL

The loss of an estimated 4 billion tons of soil from land in the United States each year (more than ten times the amount of material excavated during the construction of the Panama Canal) affects many people, but primarily the landowner. It is estimated that 3 billion tons of this total are lost from agricultural and forested lands.

Figure 2.1 shows a typical scene on a Missouri farm following a rainfall of 2.15 inches, with a maximum intensity of 4.5 inches per hour for a fifteen-minute period during the eighty-minute storm. This field was planted to corn on May 7 and the rain occurred on June 4. The

Figure 2.1. Erosion resulting from 2.15 inches of rain falling in 80 minutes.

slope on the left side of the field was not terraced and a soil loss of 88 tons per acre resulted. The slope on the right side of the field was terraced and a soil loss of 5 tons per acre resulted. During extreme rains, such as occurred in southwest Iowa in May 1950, soil losses up to 250 tons per acre may occur.

Some of the effects of loss of soil by erosion are discussed below:

Loss in production potential. The loss of 4 billion tons of soil each year is equivalent to the loss of 7 inches of soil from 4 million acres. The effect of this loss of soil on production varies, depending upon the type of soil and the depth of the topsoil. Some soils when seriously eroded are virtually useless for crop production. All soils suffer a decline in productivity. A number of studies on the effect of depth of topsoil on corn yields have been made in the corn belt states. A summation of these data indicates the following reduction in yield due to erosion (17): 2 inches of topsoil eroded result in a 15 percent reduction in yield; 4 inches—22 percent; 6 inches—30 percent; 8 inches—41 percent; 10 inches—57 percent; and 12 inches—75 percent.

Loss of nutrients needed for crop production. It is estimated that the 3 billion tons of soil eroded from our farms and forest land each year have an average analysis of 0.10 percent nitrogen (N), 0.15 percent phosphorus (P_2O_5), and 1.50 percent potassium (K_2O). This means that more than 50 million tons of plant nutrients are lost by erosion each year (18).

It is difficult to place a value on the plant nutrients and organic matter that are removed from the land by soil erosion. However if it were necessary to replace the amount removed in a ton of eroded soil at a cost of $0.07, $0.09, and $0.04 per pound for N, P_2O_5 and K_2O respectively, the cost would be approximately $1.70 per ton. The cost of replacing these nutrients in the 4 billion tons of soil that are eroded from land in the United States each year would be $6.8 billion. To replace the nutrients lost during a storm such as shown in Figure 2.1 would cost approximately $150 per acre.

Reduction in quality of crop produced. As nutrients are eroded from the soil, not only are crop yields reduced but the crops grown are lower in quality and may actually be lacking in certain nutrients (1).

Reduction of the infiltration rate and water-holding capacity of the soil. On most soils the subsoil is lower in organic matter and is not as permeable as the topsoil. As the topsoil is eroded, the subsoil does not absorb the rainfall as rapidly; consequently there will be more runoff and less water available for crop production. Also, as the topsoil is eroded, there will be less soil available for moisture storage.

In 1941 C. M. Woodruff of the Agronomy Department, University of Missouri at Columbia, removed all the topsoil from a plot on a Mexico silt loam soil. The corn yields received from this plot, as compared to a plot on a similar soil that had only slight erosion, are given in Table 2.1. Each plot was given full soil treatment so that fertility would not be a limiting factor in production.

TABLE 2.1. Corn Yields from Severely Eroded and Slightly
Eroded Plots on a Mexico Silt Loam Soil

| Year | Corn Yields (bu per acre) | |
	Severe erosion	Slight erosion
1950	67	127
1951	38	109
1952	28	56
1953	Failure	73
1954	Not harvested	
1955	76	76
1956	32	98
1957	32	65
1958	19	100
1959	52	35
1960	45	74
1961	2	99
1962	51	100
1963	70	83
1964	9	90
1965	91	128
1966	65	66
1967	86	118
1968	97	150
Average	48	92

When the rainfall was fairly adequate, not too intense, and well distributed throughout the growing season as was the case in 1955 and 1966, there was little difference in the yield from the two plots. However in years such as 1953 and 1961 when the rainfall was more intense and not as well distributed, there was an extreme difference in the resulting yields. The average yield for the period of study was 44 bushels per acre greater on the slightly eroded plot, which indicates the value of soil conservation and the resulting increase in moisture, in increasing crop production.

Deposition on fertile soil. The deposition of eroded material on fertile soils reduces their productivity and damages or destroys crops growing on the land (Fig. 2.1).

Deterioration in soil structure. As the topsoil is eroded, it becomes necessary to farm more of the subsoil which in most cases has a poorer structure. It is more difficult to prepare an adequate seedbed, and crop germination and yields are adversely affected.

More power required to till the land. The subsoil with its less desirable structure is more difficult to till and this increases the cost of producing a crop on eroded land.

Loss of cropland by gullies and stream bank erosion. As gullies become too large to cross with farm machinery, they are usually allowed to grow up in brush and trees, thus removing additional land from production. It is interesting to note on an aerial photograph of gullied land or when flying over such land, the large percentage of land that is not in production as a result of these gullies.

Division of fields by gullies. Gullies divide fields into smaller segments, increasing the time and cost required for crop production. This is quite obvious, particularly to those who have attempted to farm a severely eroded field.

Reduced income from the land. All of the above factors result in a reduction in the income-producing potential of the land, which in turn results in a decrease in land value.

SEDIMENTATION

Soil erosion, such as shown in Figure 2.1 affects many others in addition to the landowner. The sediment accumulation at the bottom of the slope in the figure consists mostly of the coarser, heavier particles that were dropped when the velocity of flow was reduced at the bottom of the slope. A much larger volume of finer material was carried on downstream in the runoff water to cause problems in stream channels, reservoirs, and harbors. Thus sedimentation resulting from soil erosion creates the following additional problems:

Reduced capacity of downstream channels and reservoirs resulting in increased flooding and reduced water supplies. An extreme example of this problem on the Yellow River in China was discussed in Chapter 1. It is obvious to anyone who has observed stream characteristics over a number of years that this problem exists to some extent on any stream where erosion occurs on the watershed. Deposition of sediment in stream channels reduces their capacity and causes them to overflow more frequently. Sediment in transport in the stream increases the volume of material carried which increases flooding.

One of the most important problems in the design and maintenance of reservoirs is the loss of storage by sedimentation. The seriousness of this problem is emphasized by the results of a survey of representative water supply reservoirs, conducted by a number of agencies in the early 1940s (2, 3). This survey indicated that more than 33 percent of the water supply reservoirs in the midwestern states will have to be supplemented or replaced before they are fifty years old. Many reservoirs are filling at a rate of 5 percent per year. This rate is even greater for some of the smaller reservoirs. Another survey indicates that about 1 million acre-feet of sediment is deposited in our artificial reservoirs each year, a volume large enough for a water supply for a city of 5.5 million people (10).

In the design of large reservoirs, a portion of the volume is reserved for sediment storage. This extra volume increases the cost of the reservoir. In planning a large reservoir in northwest Missouri, sediment storage was provided to store 7.6 inches of soil from the entire 216-square-mile drainage area. It was assumed that this would be the amount of sediment eroded from this area during the hundred-year assumed economic life of the reservoir. This sediment storage amounts to 30 percent of the usable storage to be provided in the $21,500,000 project. The cost for providing this extra storage is $6,450,000, an expenditure of $47 for each of the 138,240 acres in the drainage area. If adequate erosion control practices were provided on the watershed, this expenditure for sediment storage could be reduced a significant amount.

Many smaller ponds and reservoirs, the design of which has not included a volume for sediment storage, will be rendered essentially useless, and many will become a nuisance in a relatively short period of time. A survey of 968 reservoirs which represented a good cross section of the reservoirs in the United States gave the following information on the depletion of volume by sediment (6): the average annual depletion rate for 161 reservoirs with an original storage capacity of less than 10 acre-feet was 3.41 percent; for the 228 reservoirs with capacities ranging from 10 to 100 acre-feet it was 3.17 percent; and for the 251 reservoirs with capacities ranging from 100 to 1,000 acre-feet it was 1.02 percent. The depletion rate is higher for the smaller reservoirs. The following conclusions are made on the basis of this survey: the relatively high average storage depletion rate of over 3 percent in small upland reservoirs seems excessive; if present siltation rates continue, about 20 per cent of the nation's small reservoirs will be half-filled with sediment and in many instances their utility seriously impaired in about thirty years.

Reservoirs have been and are being constructed on the most desirable sites. If these reservoirs are allowed to fill with sediment, it will be necessary to build future reservoirs on less desirable and more expensive sites.

Increased costs of obtaining a suitable water supply. Sediment pollution of streams and reservoirs increases the expense of treating these water supplies for municipalities and industry. Potable water must be free of sediment. Many industries such as food processors and textile manufacturers require such water. Large investments in the construction and operation of treatment facilities are required to remove the sediment. The greatest amount of water used by industry is for cooling purposes. The presence of sediment increases cooling costs in the following ways: (1) silt-laden water has a higher temperature than clear water, thus necessitating use of a larger quantity; (2) silt increases the wear on pumps and equipment; (3) sediment is deposited on heat exchange surfaces, requiring greater capacities; (4) deposition of sediment in sumps and storage basins increases maintenance costs.

Reduced value of land and streams as wildlife habitats and for recreational purposes. An eroded countryside and silt-laden streams are unsightly and of limited recreational value. As the soil is depleted by erosion, wildlife find it more difficult to survive. The eroded soil washes into streams, lakes, reservoirs, and estuaries and destroys the habitat for the most desirable fish and wildlife species. Suspended sediment in the water impairs the dissolved oxygen balance and obscures the light needed for aquatic growth, both of which are detrimental to fish life. Heavier sediment particles blanket fish nesting and spawning areas and cover food supplies.

Increased cost of maintaining navigable channels and harbors. The problem of dredging and maintaining channels is of great magnitude, both as to the volume of material involved and the total sum spent in handling and disposing of the material. It is estimated that the volume of material dredged and excavated annually from streams, estuaries, and harbors exceeds one-half billion cubic yards (14). Compare this to the 211 million cubic yards that were excavated from the Panama Canal.

Decreased potential for water power. As reservoirs are filled with sediment, their storage volume is decreased, with a resulting decrease in potential for power production. In addition, sediment-laden water, in passing through power plants, increases wear on turbines.

Reduced carrying capacity and increased cost of maintenance of irrigation and drainage systems. One of the reasons for the downfall of civilizations which depended on irrigated agriculture in the past was severe erosion on the surrounding hill land. Today irrigation and drainage districts go to great expense to build structures to prevent sediment from clogging their ditches. Even greater expense is involved in removal of sediment from those ditches not protected from sedimentation.

Drainage problems and reduced productivity resulting from deposition of eroded material on productive land. In the earlier stages of agricultural development, much of the material deposited on the flood plains was very fertile and actually increased their productivity. With the acceleration of erosion on the uplands, much of the material now being eroded is relatively infertile subsoil and may reduce the productivity of the floodplain when deposited. In many instances, sand and gravel are deposited on the floodplain during floods, thus seriously limiting production. In some cases, this is plowed under with huge plows, some having moldboards as much as 8 feet tall pulled by four large track-type tractors having in excess of 500 horsepower—a very expensive operation.

Swamping is caused by the development of natural levees along stream channels by deposition of eroded material as the streams overflow their banks. These levees prevent flood water from returning to the channel, thereby causing a wet or swampy condition after the stream flow has returned to normal. These levees also interfere with the drainage of the bottomland following normal rainfall.

Increased costs of maintaining roads and highways. Following a hard rain it is not unusual to find road ditches and culverts blocked by sediment. Large expenditures are required to clear the debris and repair the damage resulting from the overflow.

Increased damage to flooded cities and homes. Unless one has attempted to clear the sediment from homes and businesses following a flood, it is difficult to realize the extent of the problem. Not only is it expensive, but the work must be accomplished under very undesirable conditions.

LOSS OF WATER

As a soil is eroded an increase in surface runoff from that soil results from a reduction in the infiltration rate and in the water-holding capacity. This loss of water will result in the following:

Decreased crop yields and more extensive drought damage. In areas where a large proportion of the rainfall occurs during the growing season, the loss of water because of a seriously eroded soil makes a significant difference in crop yields.

Increased flood damage. As erosion progresses, the infiltration and water-holding capacity of the soil is reduced and a higher percentage of the rainfall escapes in the form of runoff, increasing the flood hazard.

When most people think of floods, they have a vision of overflowing rivers, inundated cities, mud filled homes, sickness, and disease. These catastrophies cause much human suffering and deserve the attention of everyone. They are spectacular and make the headlines. Fortunately they occur at relatively infrequent intervals.

There is another type of flooding that is less spectacular and rarely causes loss of life, but it does result in economic losses greater than those sustained from floods on major rivers. This is the flooding within the tributary headwaters of the major rivers. These floods often escape our attention, but they occur on many small tributaries, often several times each year, and the total number runs into the thousands. Damage occurring on streams with drainage areas less than 250,000 acres, or approximately 400 square miles, is called upstream flood damage. A survey of rivers and streams in the conterminous United States indicates that approximately five thousand rivers have drainage areas larger than this, but some two million rivers and streams have drainage areas less than this (11). Recently, intensive efforts have been devoted to the evaluation of upstream flood damages that occur. Table 2.2 gives a summary of the annual upstream flood damage that occurs in the conterminous United States (8).

It has been estimated that the average annual flood damage in downstream areas (on streams with drainage areas greater than 250,000 acres) is approximately $550 million. Thus the flood damage on the headwater streams is approximately twice that on the major rivers.

In another study of flood damage on a typical watershed covering 10,000 square miles, the amounts of damage to different parts of the watershed were reported, each as a percentage of the total (5). City damage was reported at 15 percent of total damage, damage to the main valley—6 percent, to principal tributaries—4 percent, and to headwaters —75 percent.

TABLE 2.2. Estimated Average Annual Upstream Flood Damage in the Conterminous United States

Types of Damage	Annual Cost	Percentage of Total Cost
Crops and pasture	$459,512,000	44.9
Floodplain scour	39,972,000	3.4
Stream bank erosion	7,083,000	0.7
Gully erosion	78,397,000	7.7
Other erosion	4,082,000	0.4
Other agricultural	77,606,000	7.6
Sediment	87,652,000	8.6
Nonagricultural	182,871,000	17.9
Indirect	90,377,000	8.8
Total	$1,022,552,000	100.0

Source: E. C. Ford. Upstream flood damage. *J. Soil Water Conserv.*, vol. 19, November–December 1964.

Because a major part of flood damage occurs in the headwater
streams, added emphasis should be given to the protection of watersheds and the reduction of floods in those areas. The installation of
erosion control practices on these watersheds would be a major factor
in the reduction of flood damage for two reasons. First and most important, the soil would be retained on the watersheds, and the stream
would not be clogged with erosion debris and thus would carry a greater
amount of water without flooding. In addition the erosion control practices would reduce the peak rate of runoff expected from a given storm
on the smaller watersheds, further reducing the expected flood damage
there.

The Water Quality Act of 1965, the Water Pollution Control Act
of 1966, and various proposed or adopted state laws and standards are
evidence of the desires of the people of the United States for a better
quality environment. Since sediment is by far the greatest pollutant in
terms of volume, there will be increased emphasis on the need for a
reduction in sediment pollution.

POLLUTION

We have become accustomed to thinking of erosion as an action
that destroys good cropland and produces sediment to fill our stream
channels and reservoirs. We must now be concerned with sediment as
a source of pollution.

Chemical fertilizers and pesticides are used in great quantities to
maintain high levels of crop production. Plant residues, animal wastes,
and in some cases pathogenic bacteria are the products of farming operations. If any of these materials are carried off the land by sediment
or in surface runoff, they add to the pollution of downstream waters.
The extent of the movement of the materials or their contribution to the
pollution problem is not known. However, fish kills in some of our

streams strongly suggest that many organic insecticides gain entry into streams via surface runoff from treated agricultural land. Nitrogen fertilizer has been blamed for contaminating water supplies and, together with phosphorus, causing eutrophication of many of our lakes and rivers (16).

Phosphates are tenaciously adsorbed by soil colloids and move from farmlands into lakes and streams through erosion of soil particles on which it is adsorbed. Thus good soil conservation practices which prevent erosion are the most effective means of controlling pollution by phosphates from agricultural lands (18).

Wadleigh (19) gives an excellent report of the contribution of agriculture and forestry to the pollution problem. The fact that it is possible for many of these pollutants to be moved from agricultural lands by erosion places an increased emphasis on the need for erosion control.

PUBLIC HEALTH

Plants, which provide most of the food for both animals and man, grow in the soil, from which they derive an essential part of needed nutrients. If the soil is deficient in certain elements, the plants will not provide proper nutrition for animals, and for man who eats both plants and animals. Therefore, human health depends in large measure on the maintenance of a fertile soil (1).

The pollution and clogging of streams and reservoirs by erosion and the formation of swampy areas is one of the principal obstacles to the effective control of malaria, encephalitis, and other mosquito transmitted diseases. The movement of pollutants into ponds, rivers, lakes, and reservoirs by erosion and surface runoff creates problems in public health if these waters are used for water supply or recreation.

LOSS OF HUMAN VALUES

The loss of human values that could accompany excessive losses of our soil and water resources is a very serious problem. This loss is apparent today in many areas where the soil has been depleted by severe erosion. The farms will no longer produce an adequate income, so the farm families have a lower standard of living. They lose the pride they once had in the ownership of land, and the land and the appearance of the farm is further neglected. Associated with this loss of pride is a loss of human dignity. The individuals concerned may become dependent on society, rather than being productive members of that society.

PROBLEMS

2.1.   Select an area with which you are familiar and list the various types of erosion present and indicate how each affects the usefulness or productivity of the area.

2.2. During the period 1930–42 soil loss was measured from terraced and un-terraced watersheds at the Conservation Experiment Station, Bethany, Missouri. These watersheds had a land slope of approximately 7 percent and were in a corn, oats, wheat, meadow rotation. The average annual soil loss from the unterraced watershed was 25 tons per acre, and from the terraced watershed 0.5 tons per acre.

    a. What should be the annual cost per acre of replacing the N, P_2O_5, and K_2O lost in the eroded soil (1) on the unterraced watershed? (2) on the terraced watershed?

    b. How many inches of soil were eroded from (1) the unterraced water-shed during this period? (2) from the terraced watershed? (Assume the weight of soil to be approximately 150 tons per acre-inch.)

2.3. A single storm in southwestern Iowa in May of 1950 resulted in a soil loss of 250 tons per acre from cornfields in the areas that were not terraced. The soil loss from terraced cornfields was 5 tons per acre. Compute the information asked for in questions 2.2 (a) and (b) for this storm.

2.4. Are the people of the United States more concerned about erosion as a source of pollution than they are about the effects of erosion on reduced pro-ductivity or increased flood damages? Give evidence to substantiate your an-swer.

2.5. The manner in which a nation manages its soil and other resources is in-fluenced by the attitudes, customs, and traditions of the people and by the social, religious, and political institutions they have developed. Considering the conditions that exist in the United States today, outline a program that would result in effective erosion control on farmland in the near future. See Burton and Kates (4), Ford et al. (9), and Matson et al. (12) for selected readings in this area.

REFERENCES

1. Albrecht, W. A. Nutrition via soil fertility according to the climatic pattern. Commonwealth Scientific and Industrial Research Organization, Melbourne, Australia, May 1949.
2. Brown, C. B. The control of reservoir silting. USDA Misc. Publ. 521, 1943.
3. ———. Effect of land use on reservoir siltation. *J. Am. Water Works Assoc.*, vol. 41, no. 10, October 1949.
4. Burton, Ian, and Kates, R. W. *Readings in resource management and conservation.* Selected readings from Part II, The conservation of limited resources. Chicago: Univ. Chicago Press, 1965.
5. Cook, H. L. Flood abatement by headwater measures. *Civil Eng. J.*, March 1945.
6. Dendy, F. E. Sedimentation in the nation's reservoirs. *J. Soil Water Conserv.*, vol. 23, no. 4, July–August 1968.
7. Ellison, W. D. Soil detachment hazard by raindrop splash. *Agr. Eng. J.*, vol. 28, no. 5, May 1947.
8. Ford, E. C. Upstream flood damage. *J. Soil Water Conserv.*, vol. 19, Novem-ber–December 1964.
9. Ford, E. C., Cowan, W. L., and Holtan, H. N. Floods and a program to alleviate them. In *Water*, USDA Yearbook of Agriculture, 1955.
10. Glymph, L. M., and Storey, H. C. Sediment—Its consequences and control. Am. Assoc. Advan. Sci. Publ. 85, Washington, D.C.
11. Leopold, Luna B. Rivers. *Am. Sci.*, vol. 50, no. 4, December 1962.

12. Matson, H. O., Heard, W. L., Lamp, G. E., and Ilch, D. M. The possibilities of land treatment in flood prevention. In *Water*, USDA Yearbook of Agriculture, 1955.
13. Osborn, Ben. How rainfall and runoff erode the soil. In *Water*, USDA Yearbook of Agriculture, 1955.
14. Roehl, John W. Cost of dredging and maintaining channels. *J. Soil Water Conserv.*, vol. 20, no. 4, July–August 1965.
15. Schwab, G. O., Frevert, R. K., Edminster, T. W., and Barnes, K. K. *Soil and water conservation engineering*. New York: Wiley, 1966.
16. Sievers, D. M., Lentz, G. L., and Beasley, R. P. Movement of agricultural fertilizers and organic insecticides in surface runoff. Trans. Am. Soc. Agr. Eng., vol. 13, no. 3, 1970.
17. Stallings, J. H. *Soil conservation*. Englewood Cliffs, N.J.: Prentice-Hall, 1964.
18. Taylor, A. W. Phosphorus and water pollution. *J. Soil Water Conserv.*, vol. 23, no. 6, November–December 1966.
19. Wadleigh, C. H. Wastes in relation to agriculture and forestry. USDA Misc. Publ. 1065, March 1968.

3 ∽ Wind Erosion

Wind is an important geological agent influencing soil erosion. It detaches, transports, deposits, and mixes soil and is a factor in soil formation as well as soil erosion (11).

SOIL EROSION BY WIND becomes a serious problem when natural vegetation is removed or depleted. Animals, insects, diseases, and men have contributed to the removal or depletion of vegetation, leaving the soil susceptible to the force of the wind. Land that should have remained in natural vegetation has been placed in cultivation. Land that is suitable for cultivation under proper management has suffered from wind erosion as a result of untimely cultivation, use of improper equipment, and burning of crop residues. Poor management of land and animals has contributed to wind erosion on range and pasture land.

Soil erosion by wind is usually considered to be of serious consequence only in arid and semiarid regions; it may occur wherever soil, vegetation, and climatic conditions are conducive (19). These conditions exist when the soil is loose, dry, and finely granulated; the soil surface is reasonably smooth, and vegetative cover is nonexistent or sparse; there are large fields with no obstructions to reduce the force of the wind; and the wind is sufficiently strong to initiate soil movement.

Areas of agricultural land subject to wind erosion are located in North Africa, the Near East, central, south, and eastern Asia, Australia, southern South America, and parts of North America (9, 19). It is the dominant problem on about 70 million acres of land in the United States—55 million acres of cropland, 9 million acres of rangeland, and 6 million acres of "other" land. The agricultural areas most subject to wind erosion are the Columbia River Basin, the muck and sandy soils around the Great Lakes, the Gulf and Atlantic seaboards, and the Great Plains.

The Great Plains, because of its climatic characteristics of low and

25

variable precipitation, high frequency of drought, high temperatures and evaporation rates, and variable high wind velocities, has a long history of wind erosion. From 1854 to 1964, there were thirteen periods when major dust storms occurred. The two worst periods were in 1936–37, when there were 120 storms reported at Dodge City, Kansas, and in 1955–56, when there were 40 storms reported at this location. Since 1956, no major storms have occurred; however during this period an average of 2.7 million acres per year have been damaged by wind erosion.

PROBLEMS ASSOCIATED WITH WIND EROSION

Many of the problems associated with the loss of soil by water erosion also apply to wind erosion. Dust storms associated with severe wind erosion cause additional problems. People and animals suffer from dust inhalation and respiratory and eye infections. Fences, buildings, roads, railroads, ditches, grass, shrubs, and trees are damaged by the drifting soil. Dust storms pollute the atmosphere—as much as 1,290 tons of dust per cubic mile of air have been measured during a severe dust storm (8) (Fig. 3.1).

Wind erosion not serious enough to be classed as a dust storm may also result in significant damage. Seedling crops may be destroyed by the blowing soil. The yield and quality of crops may be reduced as a result of the abrasive damage. Sufficient soil is often removed to expose the plant roots or ungerminated seed, thus reducing or destroying the stand.

The most serious damage is the change in soil texture, physical

Figure 3.1. Severe dust storm. (Courtesy Agricultural Research Service)

condition, and fertility caused by wind erosion. The finer soil particles and organic matter are carried away, leaving the coarse, less productive soil particles. This sorting action of the wind not only removes the most important materials, from the standpoint of productivity and water retention, but leaves a sandier, more erodible soil than the original. As wind erosion continues, a soil condition is created wherein plant growth is restricted and erodibility increased. The sands may begin to drift and form unstable dunes which encroach on more productive land. Large areas of productive land have been ruined in this manner.

THE PROCESS OF WIND EROSION

The erosion process consists of three distinct phases involving soil particles—initiation of movement, transportation, and deposition—and is initiated when a strong, turbulent surface wind generates sufficient energy to overcome gravity.

INITIATION OF MOVEMENT

Movement of soil particles is caused by wind forces exerted against the surface of the ground. The average forward velocity of the wind near the ground increases exponentially with height above the ground surface. The change in velocity with height is known as the velocity gradient. It is this gradient which determines the shear stress or drag force exerted (2, 9, 19).

At some point near the surface of the ground the wind velocity is zero. The zero velocity is usually somewhat above the average roughness elements of the surface. The higher the roughness elements of the ground, or the taller and less air-permeable the vegetative cover, the higher the level at which zero velocity is found. Above this level the average velocity increases rapidly at first, and then less rapidly as the height increases. In this zone the wind is turbulent and is characterized by eddies moving at variable velocities and in all directions (24).

Soil particles or other projections on the surface protruding into this layer of turbulent wind absorb most of the force exerted on the surface. If they are large enough or firmly attached to other particles, they may resist the force of the wind. If however they are not attached and are not too heavy, the wind may lift them from the soil surface and initiate soil movement.

There are three types of pressure applied to soil grains as wind passes over them: positive, or impact, pressure on the windward side; negative, or drag, pressure on the leeward side; and negative, or lifting, pressure on the top side (6).

The minimum wind velocity required to initiate movement of the most erodible soil particles, those about 0.004 inch (0.1 mm) in diameter, is about 10 miles per hour at 1 foot above the soil surface. These particles have a size and weight relationship which is most conducive to initiation of movement. They protrude to a sufficient height into the turbulent layer to absorb an appreciable force, yet are light enough to be easily moved. Smaller particles exhibit more cohesion and do not protrude to as great a height into the turbulent layer, hence a stronger wind is required to initiate movement. Larger particles protrude a greater distance into the turbulent layer but they are heavier, so a stronger wind is required to initiate movement. For dune sand having a mixture of different sized particles, the minimum wind velocity at 1 foot of height required to initiate soil movement is usually assumed to be about 13 miles per hour.

TRANSPORTATION

After movement of soil particles is initiated, they are carried by saltation, surface creep, or suspension, depending upon their size in relation to the velocity and turbulence of the wind.

Most particles are moved by saltation, which consists of a bouncing or jumping of the particles over the surface. The force of the wind lifts the particles into the air. They rise almost vertically, rotating at several hundred revolutions per second, travel ten to fifteen times their height of rise, and return to the surface at forward and downward angles of 6°–12° (23). Those particles in the size range of 0.004–0.02 inch (0.1–0.5 mm) are usually moved by saltation.

Surface creep is rolling or sliding of the soil particles along the surface. They are too heavy to be lifted by the wind and are pushed or rolled by the impact of the smaller particles in saltation. Those particles in the size range of 0.02–0.04 inch (0.5–1.0 mm) are usually moved by surface creep. The rate of soil movement in surface creep and saltation is proportional to the cube of the wind velocity (24).

Particles smaller than 0.004 inch (0.1 mm) may be moved by suspension. Movement of these fine particles is usually initiated by the impact of particles in saltation. Once they enter the turbulent air layers, they may be lifted high into the air by upward eddy currents and may be carried many miles. The greatest amount of soil is moved by saltation and surface creep, but that moved by suspension is the most spectacular and the most easily recognized (Fig. 3.1).

Once soil particles are loosened and movement is initiated, the impact of particles in saltation severely abrades the soil surface, damages vegetation, and breaks down clods and surface crusts. Thus the saltating particles not only initiate movement of both larger and smaller particles, but they also break clods up into erodible sizes and reduce the effectiveness of vegetative cover. The larger the area, the greater the number of times individual particles strike the surface as they move with the wind. Consequently an increasing number of particles are set in motion as the distance downwind increases. Erodible soil particles accumulate on the surface, causing progressively greater concentration of impacting grains until a maximum amount of soil movement that a given wind can sustain is reached. This effect is referred to as avalanching.

DEPOSITION

The soil particles moved during the wind erosion process are deposited at a new location when the wind subsides or when surface obstructions alter the velocity distribution and turbulent structure (3, 17, 20). This deposition may be considered beneficial if it adds to the fertility of an existing soil or detrimental if deposited on more fertile soil and on cities and homes. Wind-transported sediment is a major factor in geological changes constantly occurring over the land surface (11, 18). Man has greatly accelerated these changes since he began cultivating the soil.

FACTORS AFFECTING WIND EROSION

The major factors affecting wind erosion are climate, soil, surface roughness, vegetation, and length of erodible surface along the prevailing wind direction.

CLIMATE

Climatic factors affecting wind erosion are precipitation, temperature, humidity, and wind. A moist soil is more stable than a dry one because of the cohesion of the water film surrounding the particles. Thus the moisture content of the soil at the time when strong, turbulent winds occur is a major variable influencing wind erosion. The amount and distribution of precipitation determines moisture inputs. The temperature, humidity, and wind affect evaporation and transpiration and determine moisture withdrawal.

Intense rains consolidate and crust the soil surface, increasing its resistance to erosion. Alternate wetting and drying and freezing and thawing tend to disintegrate the crust and clods, causing the soil to be more susceptible to wind erosion. Winds provide the force to move the soil particles. The principal characteristics of wind influencing the amount of soil movement are velocity, turbulence, direction, and duration.

SOIL CHARACTERISTICS

Soil characteristics affecting its erodibility are texture, density, and structural stability (5).

The size and density of a soil particle determine its weight and its erodibility. The lighter soil particles are more erodible than heavier ones, but only if they are large enough to protrude into the turbulent layer of the wind far enough to absorb an appreciable force. The divid-

ing line between erodible and nonerodible particles is not distinct, for
it varies with the velocity of the wind and the size range and density of
the particles. For soil particles with a density of 2.65, those with a diameter of about 0.004 inch (0.1 mm) are considered the most erodible.
The wind velocity required to initiate movement of these particles is
about 10 miles per hour at 1 foot above the soil surface. Relatively few
particles greater than 0.02 inch (0.5 mm) in diameter are moved by the
more common erosive winds; however, particles as large as 0.08 inch
(2.0 mm) in diameter may be moved by exceedingly high winds. Particles less than 0.0008 inch (0.02 mm) in diameter are highly resistant to
movement by direct force of the wind, particularly if the soil surface is
smooth. Also, because of their cohesiveness they hinder the movement
of larger particles mixed with them. If a soil surface composed of fine
particles is roughened so that the surface is projected into the turbulent
layer of the wind, the particles erode readily until the surface is leveled
by the wind.

SURFACE ROUGHNESS

A roughened soil surface reduces the velocity of the wind and the
forces tending to move soil particles. Surface roughness may be increased
by clods and by ridges and depressions formed by tillage implements.
While the general effect of surface roughness is to reduce wind erosion,
it also increases wind turbulence and exposes smaller areas to greater
wind forces. Therefore if the variations in height of the roughened surface are too great, benefits derived may be substantially reduced. Optimum variations in the height of the roughened surface for most effective wind erosion control are from 2 to 5 inches (7, 21). Stable clods on
the surface of the soil increase its resistance to erosion and reduce the
erosive force of the wind.

VEGETATION

Vegetation reduces the wind velocity at the soil surface and absorbs
much of the force exerted by the wind. Soil particles are also trapped
by the vegetation, thus preventing avalanching of soil particles downwind. Standing vegetation is most effective; however, vegetative residue,
if anchored in the soil, is beneficial in reducing wind erosion. The importance of vegetative protection on the land cannot be overemphasized.
It is nature's way and is the "cardinal" or "golden" rule of wind erosion
control (9).

LENGTH OF ERODING SURFACE

Avalanching has been described as the increase in rate of erosion
with distance downwind over an unprotected eroding area. The rate
increases from zero on the downwind edge to a maximum a given wind
can sustain, provided that the eroding area is of sufficient length. There-

fore obstructions that break up the length of the eroding surface reduce wind erosion. Crop strips accomplish this, as do barriers.

WIND EROSION EQUATION

A wind erosion equation has been developed to evaluate the total influence of the major variables influencing wind erosion (9, 22). The equation is $E = f(I, K, C, L, V)$, where E is the soil loss in tons per acre per year that could occur from a given field, I is a soil-erodibility index related to cloddiness, K is soil roughness, C is a climatic factor related to wind velocity and soil moisture, L is the field length along the prevailing wind direction, and V is the equivalent quantity of vegetative cover. This equation can be used to determine the potential erodibility of farm fields or to design erosion control practices to reduce erosion to a tolerable amount. Charts, tables, and maps giving information relating to the magnitude of the wind erosion forces in different parts of the United States and providing information necessary for the solution of the above equation have been published (15). A computer solution for the equation is also available, permitting the user to apply the equation after recording necessary data relative to the field on two simple forms (16).

WIND EROSION CONTROL

A study of the mechanics of wind erosion and the factors influencing it suggest three basic methods for its control: increasing soil stability and surface roughness; establishing and maintaining vegetation, crop residue, or other types of cover; and placing barriers perpendicular to the direction of the prevailing wind. Based on his present knowledge, man can do little to control climatic factors; however, he can influence the temperature, humidity, and the moisture supply to the soil to some extent by irrigation. He can also influence the velocity of the wind at the soil surface by creating barriers perpendicular to the prevailing wind.

SOIL STABILITY AND ROUGHNESS OF SURFACE

Man can do little to modify the texture and density of a soil. He can however influence soil stability by his management of the soil. He can influence soil moisture, organic matter, and microorganism activity to some extent. Certain petroleum and chemical products are being evaluated to determine their possible use as soil stabilizers. However at the present time his greatest influence on soil stability is by tillage practices.

The type and amount of tillage affects the consolidation of soil particles, the cloddiness, and the roughness of the soil surface. A type of tillage should be used that will prepare an adequate seedbed and control weeds. However for wind erosion control the surface must be left rough with maximum consolidation of the soil particles. Mold-board plows leave the soil surface rough and cloddy, but they turn under crop residue. Disk plows and harrows tend to mix and pulverize the soil, leaving it more susceptible to wind erosion. Field and chisel cultivators mix the residue with the surface soil and create a rough sur-face. The lister creates a ridged surface which is quite resistant to wind erosion, particularly if the ridges are perpendicular to the prevailing winds. Subsurface sweeps and blades leave a large percentage of crop residue on the surface; however, they do not produce a cloddy or roughened surface and may not kill all weeds under moist conditions. Rod weeders are effective in controlling weeds while leaving a high per-centage of the residue on the surface.

Because of the difficulty of meeting all the requirements for seedbed preparation, erosion-resistant surface, and weed control by one tillage method, the two-zone concept of tillage may have application. In this method, a narrow strip of soil is prepared to give good seed germination, and the inter-row area is left in a condition to provide maximum wind erosion control.

In addition to regular tillage required in crop production, emer-gency tillage is often required to reduce wind erosion losses when vege-tative cover is depleted by unfavorable weather conditions, improper management, or excessive grazing. The objective of emergency tillage should be the creation of a rough, cloddy surface which will reduce the force of the wind, leave the soil in a stable condition, and trap the blowing soil. Field or chisel cultivators and listers are commonly used for emergency tillage. Emergency tillage is most effective in controlling wind erosion if it is accomplished when the soil is moist and compact and if there is sufficient clay to produce stable clods (21). The direction of tillage should be perpendicular to the direction of prevailing winds (Fig. 3.2).

Only sufficient tillage should be done to prepare an adequate seed bed and control weeds. Excessive tillage pulverizes the clods and smooths the soil surface, leaving the soil more susceptible to erosion.

VEGETATION, RESIDUES, AND ARTIFICIAL COVERS

The most effective means of controlling wind erosion is to maintain a protective cover on the soil surface. Close growing crops such as grasses and small grain, once they are established, provide excellent protection. Row crops such as corn, sorghum, and cotton are less effective because of the space between rows. To be most effective, row crops should be planted with the crop rows perpendicular to the prevailing wind. Cover crops such as wheat, rye, oats, or vetch can be used to provide protection when that provided by the regular crop is not adequate. The proper sequencing in a crop rotation can be used to provide protection at times of the year when the soil is most vulnerable to wind erosion.

Figure 3.2.   Emergency tillage for wind erosion control. (Courtesy Agricultural Research Service)

A cover of grass, trees, or shrubs should be established and maintained on noncropland whenever possible. Controlled grazing and proper fertility and management practices will be necessary to maintain these covers.

Crop residues, if properly managed, provide additional protection from wind erosion after the crop is harvested and during the period when the next crop is being established. Undisturbed residues which remain in an essentially upright position provide the maximum effectiveness. Any action such as tillage, grazing, or natural decomposition which tends to flatten or cover the residue reduces its effectiveness. The amount of protection depends also on the quantity and quality of the residue. The finer-textured and denser residues such as wheat and rye give more protection for a given quantity of residue, than do coarser-textured residues such as corn and sorghum. The effectiveness increases as the quantity increases. Crops such as soybeans provide little residue and thus are quite susceptible to wind erosion.

In order to reduce wind erosion to an acceptable level, the following rates of residue are suggested: a silt loam soil with 25 percent nonerodible fractions needs 750 pounds of 12-inch standing wheat stubble or 1,500 pounds of 12-inch flattened wheat stubble per acre; a loamy sand with 25 percent nonerodible fractions needs 1,750 pounds of standing

wheat stubble or 3,500 pounds of flattened wheat stubble per acre; if
sorghum stubble is substituted for wheat stubble, the quantity is doubled
(10).

Tillage implements and management practices must be selected to
maintain the maximum amount of crop residue to protect the land until
the following crop provides an adequate cover. Artificial covers such
as gravel, wood cellulose fiber, and cotton gin trash, and surface films
and binders such as petroleum-based binders, as well as various resin and
latex emulsions may be used to control wind erosion on limited areas
such as highway and airport right-of-ways (Fig. 3.3).

Vegetation, residues, and other covers of adequate quality give
protection from wind erosion. These measures sometimes fail because
high-velocity winds frequently occur concurrently with seedbed preparation and early growth of crops seeded in fall and spring. Properly designed barriers may be used to provide additional protection during
these periods (13).

BARRIERS

Any obstruction which breaks up or reduces the length of an eroding
surface reduces wind erosion. Wind barriers such as trees, crops, and
snow fences placed perpendicular to the direction of the wind deflect
the wind stream, reduce the wind velocity near the ground surface downwind from the barrier, and trap blowing soil particles. The height,
shape, porosity, length, and location of the barrier relative to the direction of the wind are important factors in their effectiveness.

When the wind blows at right angles to a typical tree shelterbelt,

*Figure 3.3. Stabilizing soil with a mixture of straw and asphalt. (Courtesy
Agricultural Research Service)*

wind velocity is reduced by 70 to 80 percent near the belt. The protected area is usually somewhat triangular in shape; thus the area protected increases with the length of the barrier. At a distance equal to twenty times the height of belt, the velocity is reduced by 20 percent, and at a distance equal to thirty to forty belt heights leeward, no reduction in velocity occurs. The higher the average wind velocity, the closer the belts or other barriers should be spaced to prevent the soil from blowing (9, 14). The shape of the surface of the barrier in contact with the wind influences its effectiveness. A vertical or very abrupt surface provides less protection than a sloping outer surface. A tree windbreak consisting of several species, with the shorter trees on the outer surface, is more effective than one consisting of one species (9).

Dense, relatively nonporous barriers result in a large reduction in wind velocity near the barrier. More porous barriers do not provide as great a reduction near the barrier but are effective for a greater distance downwind.

In addition to contolling wind erosion, barriers have the potential for improving the soil-water environment of crops by trapping and distributing snow over the sheltered area for subsequent soil storage. In the absence of vegetation or other barriers, fields are swept bare. Snow that would otherwise provide cover and a source of soil water terminates in gullies, fence rows, or other obstacles that reduce wind velocity and cause deposition. The distribution of snow in the area to be cropped must be relatively uniform so that recharge of soil moisture is in turn uniform. This would indicate that short, closely spaced barriers should perform better than tall, widely spaced barriers (13).

By reducing wind velocities, barriers decrease the evaporation potential between barriers, and provide a more favorable environment for the plant (13).

Crops. Strips or rows of erosion-resistant crops are frequently alternated with strips of erosion-susceptible crops or fallow to reduce the length of the eroding surface (Figs. 3.4 and 3.5). Crop strips are not an effective barrier in reducing wind velocity; however, they do trap soil particles and reduce field length and soil avalanching.

The actual width of strips varies with factors affecting field erodibility such as soil texture, cloddiness, and roughness, and wind velocity and direction. The width may range from a single row of grain sorghum or corn to several rods (4, 12). Erosion-resistant crops are small grains and other closely seeded crops that produce a crop cover rapidly. Erosion-susceptible crops are cotton, tobacco, sugarbeets, peas, beans, peanuts, asparagus, and most truck crops. In the Great Plains strip-cropping may consist of alternate strips of wheat and grain sorghum, or fallow wheat stubble and newly seeded wheat. In vegetable-growing areas alternate strips of rye and vegetables are commonly used. Crop strips should be planted at right angles to the prevailing winds.

Figure 3.4.   Grass barriers to control wind erosion. (Courtesy Agricultural Research Service)

Trees. Windbreaks, or shelterbelts, composed of several rows of trees have been used as barriers. However they have definite limitations as a general method of wind erosion control. They take up space, interfere with the operation of machinery, and compete with crops for moisture and nutrients (9).

Tree and shrub shelterbelts should be spaced at intervals of 350 to 450 feet on highly erosive soils such as sands, and from 500 to 650 feet on moderately erosive soils such as loams (9, 14).

Figure 3.5.   Strip-cropping to control wind erosion. (Courtesy Agricultural Research Service)

*Figure 3.6. Tree windbreaks and tree barriers used to control wind erosion.
(Courtesy Agricultural Research Service)*

Windbreaks are very beneficial around farmsteads, watering sites, corrals, and similar areas. They not only reduce wind erosion in these areas, but they provide protection from winds for animals and humans. They also have aesthetic qualities which may be an asset (19).

In the past multiple-row tree barriers have been used as field barriers in the Great Plains. More recently multiple-row barriers have given way to single-row barriers because of the large proportion of land occupied by multiple-row barriers, the nonuniform distribution of snow, and recent research and experience showing the comparatively high performance of single-row barriers. Design objectives of the single-row barriers have with time evolved from the dense, narrow, within-row spacing of trees to the more permeable wider-spaced plantings. Concentrated water erosion from deep snow drifts in the immediate vicinity of the dense barriers and the resulting delay of spring field operations, plus the effectiveness of the single-row barrier for a greater distance downwind have prompted this evolution (Fig. 3.6) (13).

Structures. Artificial barriers such as snow fences and board walls are quite expensive. They are used to a limited extent for wind erosion control around farmsteads, corrals, and similar areas.

REFERENCES

1. Anderson, D. T., and committee. Soil erosion by wind—Cause, damage, control. Can. Dept. Agr. Publ. 1266, 1966.

2. Bagnold, R. A. *The physics of blown sand and desert dunes.* New York: William Morrow, 1943.

3. Chepil, W. S. Sedimentary characteristics of dust storms. I. Sorting of wind eroded material. *Am. J. Sci.,* vol. 255, 1957.

4. ———. Width of field strips to control wind erosion. Kans. Agr. Expt. Sta. Tech. Bull. 92, 1957.

5. ———. Soil conditions that influence wind erosion. USDA Tech. Bull. 1185, 1958.

6. ———. Equilibrium of soil grains at the threshold of movement by wind. Proc. Soil Sci. Soc. Am., vol. 23, 1959.

7. Chepil, W. S., and Milne, R. A. Wind erosion in relation to roughness of the surface. *Soil Sci.,* vol. 52, 1941.

8. Chepil, W. S., and Woodruff, N. P. Sedimentary characteristics of dust storms. II. Visibility and dust concentration. *Am. J. Sci.,* vol. 255, 1957.

9. ———. The physics of wind erosion and its control. *Advan. Agron.,* vol. 15, 1963.

10. Chepil, W. S., Woodruff, N. P., and Siddoway, G. H. How to control soil blowing. USDA Farmers Bull. 2169, 1961.

11. Free, E. E. The movement of soil material by wind. USDA Bureau of Soils Bull. 68, 1911.

12. Hill, R. G. Wind erosion control on upland soils. Farm Sci. Ser. Mich. State Univ. Ext. Bull. 525, 1966.

13. Siddoway, F. H. Barriers for wind erosion control and water conservation. *J. Soil Water Conserv.,* vol. 25, no. 5, September–October 1970.

14. Skidmore, E. L. Modifying the microclimate with wind barriers. Great Plains Agr. Council Publ. 34, vol. 1, 1969.

15. Skidmore, E. L., and Woodruff, N. P. Wind erosion forces in the United States and their use in predicting soil loss. USDA Agr. Handbook 346, April 1968.

16. Skidmore, E. L., Fisher, P. S., and Woodruff, N. P. Wind erosion equation: Computer solution and application. Proc. Soil Sci. Am., vol. 34, no. 6, 1970.

17. Smith, R. M., Twiss, P. C., Krauss, R. K., and Brown, M. J. Dust deposition in relation to site, season, and climatic variables. Proc. Soil Sci. Soc. Am., vol. 34, 1970.

18. Udden, J. S. Erosion, transportation, and sedimentation performed by the atmosphere. *J. Geol.,* vol. 2, 1894.

19. United Nations. Soil erosion by wind and measures for its control on agricultural lands. FAO Dev. Paper 71, Rome, 1960.

20. Waggoner, P. E., and Bingham, C. Depth of loess and distance from source. *Soil Sci.,* vol. 92, 1961.

21. Woodruff, N. P., and Lyles, L. Tillage and land modification to control wind erosion. In Tillage for greater crop production, Am. Soc. Agr. Eng. Publ. Proc–168, 1967.

22. Woodruff, N. P., and Siddoway, F. H. A wind erosion equation. Proc. Soil Sci. Soc. Am., vol. 29, no. 5, 1965.

23. Zingg, A. W. Some characteristics of Aeolian sand movement by saltation process. Editions du Centre National de la Recherche Scientifique 13, Quai Anatole France, Paris, 1953.

24. Zingg, A. W., Chepil, W. S., and Woodruff, N. P. Sediment transportation mechanics: Wind erosion and transportation. *J. Hydraul. Div.,* Proc. Am. Soc. Civ. Eng., vol. 91, paper 4261, 1965.

4 ～ Soil-Loss Prediction Equation

To develop an effective erosion control program, it is necessary to evaluate the factors affecting erosion and practices for its contol.

T HE GREATEST AMOUNT of erosion occurs on our nation's farmland and practices have been developed to reduce it. Erosion on nonfarmland, although not as extensive, is a serious problem in many areas and some farmland practices can be used in reducing it. In planning his farm layout, the farmer must select the combination of agronomic and mechanical practices that will best conserve his soil and provide him with an efficient business operation. The income-producing enterprises require certain types of feed in amounts that are very nearly uniform from year to year. Good management requires the use of a cropping system that will furnish this feed and supporting practices that will conserve the soil. It is not possible to avoid all loss from erosion, but there is a point where the soil loss will be sufficiently small that crop production can be carried on and the productivity of the soil maintained or perhaps increased through the years.

Specific guidelines are presented here to help select practices best suited to the particular needs of each farm. The figures and tables and much of the text of this chapter have been taken from Wischmeier and Smith (1).

FACTORS AFFECTING SOIL LOSS

Scientific farm planning for soil and water conservation requires knowledge of the relations between those factors that cause loss of soil and water and those practices that help reduce such losses. Since 1930 controlled studies on experimental plots and small watersheds have supplied much valuable information on these relationships. Most of the basic information on runoff and soil loss obtained from studies in the United States were assembled at the Runoff and Soil-Loss Data Center of Agricultural Research Service located at Purdue University. See Wischmeier and Smith (1). Analysis of this data has resulted in an improved soil-loss prediction, or erosion, equation:

$$A = RKLSCP$$

where $A =$ soil loss in tons per acre
$R =$ rainfall factor
$K =$ soil-erodibility factor
$L =$ slope-length factor
$S =$ slope-gradient factor
$C =$ cropping-management factor
$P =$ erosion control practice factor

RAINFALL (R)

Development of the erosion equation necessitated determination of specific rainfall characteristics closely correlated to erosion and which could be measured and evaluated. Research data showed that when factors other than rainfall were held constant, the soil losses from cultivated fields were directly proportional to the product value of two rainstorm characteristics—total kinetic energy of the storm times its maximum thirty-minute intensity. This erosion index *(EI)* value for any particular storm can be computed from a recording-rain-gage chart and tables giving rainfall energies. The sum of the computed storm erosion index values for a given period is a numerical measure of the erosivity of all the rainfall within that period. The rainfall erosion index at a particular location is the longtime average yearly total of the storm erosion index values. The lines in Figure 4.1 with the same erosion index value are called iso-erodents (which implies equally erosive average annual rainfall). The rainfall erosion index values for the mountainous states west of the 104th meridian were not included because of the sporadic rainfall pattern of this area.

When the soil-loss equation is used to estimate average annual soil loss, the factor R is the average annual value of the erosion index and is obtained from Figure 4.1. If soil losses other than average annual losses are desired, then values other than average annual values of the erosion index must be substituted for R in the equation. For example, the quantity of soil loss that will be exceeded one year in five, on the average, may be estimated by assigning to R the 20 percent probability of the erosion index. The 50 percent, 20 percent, and 5 percent probability values of the erosion index at key locations are given in Table 4.1. The range of the erosion index observed at these locations during twenty-two years of record is also given in Table 4.1.

The approximate amount of soil loss from a single storm that will

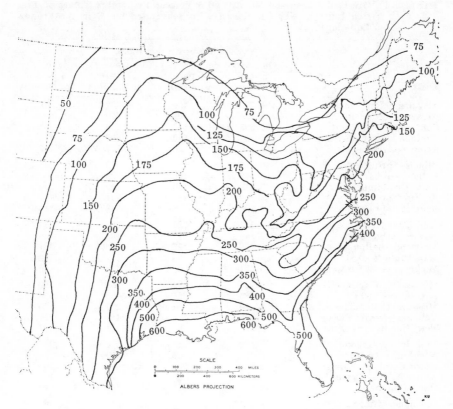

Figure 4.1. Average annual values of the rainfall factor, R.

probably be exceeded once in one, two, five, ten, or twenty years can be determined by using the appropriate value of the erosion index as *R* in the soil-loss equation (Table 4.2). For the calculation, the value of *C* should be determined for the specific conditions existing in the field at the time of the storm.

SOIL ERODIBILITY (K)

Some soils erode more readily than others even when rainfall, slope, vegetative cover, and management practices are the same. This difference, due to the properties of the soil itself, is referred to as soil erodibility. Soil properties that influence erodibility by water are (1) those that affect the infiltration rate, permeability, and total water capacity and (2) those that resist the dispersion, splashing, abrasion, and transporting forces of the rainfall and runoff.

The soil properties most desirable from a crop production stand-

TABLE 4.1. Observed Range and 50, 20, and 5 Percent Probability Values of Erosion Index at Key Locations (to Be Substituted for R in the Erosion Equation)

| Location | Values of Erosion Index | | | |
	Observed 22-year range	50 percent probability	20 percent probability	5 percent probability
Alabama—Montgomery	164–780	359	482	638
Arkansas—Little Rock	103–625	308	422	569
Connecticut—Hartford	65–355	133	188	263
Florida—Jacksonville	283–900	540	693	875
Georgia—Atlanta	116–549	286	377	488
Illinois—Springfield	38–315	154	210	283
Indiana—Indianapolis	60–349	166	225	302
Iowa—Des Moines	30–319	136	198	284
Kansas—Concordia	38–569	131	241	427
Kentucky—Lexington	54–396	178	248	340
Louisiana—New Orleans	273–1366	721	1007	1384
Maine—Skowhegan	39–149	78	108	148
Maryland—Baltimore	50–388	178	263	381
Massachusetts—Boston	39–366	99	159	252
Michigan—East Lansing	35–161	86	121	166
Minnesota—Minneapolis	19–173	94	135	190
Mississippi—Vicksburg	165–786	365	493	658
Missouri—Columbia	98–419	214	297	406
Nebraska—North Platte	14–236	81	136	224
New Hampshire—Concord	52–212	91	131	187
New Jersey—Trenton	37–382	149	216	308
New York—Rochester	22–180	66	101	151
North Carolina—Raleigh	152–569	280	379	506
North Dakota—Devils Lake	21–171	56	90	142
Ohio—Columbus	45–228	113	158	216
Oklahoma—Guthrie	69–441	210	316	467
Pennsylvania—Harrisburg	48–232	105	146	199
Rhode Island—Providence	53–225	119	167	232
South Carolina—Columbia	81–461	213	298	410
South Dakota—Huron	18–145	60	91	136
Tennessee—Nashville	116–381	198	262	339
Texas—Austin	59–669	270	414	624
Vermont—Burlington	33–270	72	114	178
Virginia—Roanoke	78–283	129	176	237
West Virginia—Elkins	43–223	118	158	209
Wisconsin—Madison	38–251	118	171	245

point are not necessarily so when considering a soil's resistance to being detached and transported by erosive forces. For example, a soil that is loose, granular, and in excellent physical condition will be more severely eroded than one in poor condition if an excessive rain occurs when the soil is not protected by a vegetative cover. If a vegetative cover cannot be maintained during the time that excessive rains occur, then additional measures to protect the soil from serious erosion will be necessary.

TABLE 4.2. Expected Magnitudes of Single Storm Erosion Index Values (to Be Substituted for R in the Erosion Equation)

Location	Index Values Normally Exceeded Once				
	1 year	2 years	5 years	10 years	20 years
Alabama—Montgomery	62	86	118	145	172
Arkansas—Little Rock	41	69	115	158	211
Connecticut—Hartford	23	33	50	64	79
Florida—Jacksonville	92	123	166	201	236
Georgia—Atlanta	49	67	92	112	134
Illinois—Springfield	36	52	75	94	117
Indiana—Indianapolis	29	41	60	75	90
Iowa—Des Moines	31	45	67	86	105
Kansas—Concordia	33	53	86	116	154
Kentucky—Lexington	28	46	80	114	151
Louisiana—New Orleans	104	73	99	121	141
Maine—Skowhegan	18	27	40	51	63
Maryland—Baltimore	41	59	86	109	133
Massachusetts—Boston	17	27	43	57	73
Michigan—East Lansing	19	26	36	43	51
Minnesota—Minneapolis	25	35	51	65	78
Mississippi—Vicksburg	57	78	111	136	161
Missouri—Columbia	43	58	77	93	107
Nebraska—North Platte	25	38	59	78	99
New Hampshire—Concord	18	27	45	62	79
New Jersey—Trenton	29	48	76	102	131
New York—Rochester	13	22	38	54	75
North Carolina—Raleigh	53	77	110	137	168
North Dakota—Devils Lake	19	27	39	49	59
Ohio—Columbus	27	40	60	77	94
Oklahoma—Guthrie	47	70	105	134	163
Pennsylvania—Harrisburg	19	25	35	43	51
Rhode Island—Providence	23	34	52	68	83
South Carolina—Columbia	41	59	85	106	132
South Dakota—Huron	19	27	40	50	61
Tennessee—Nashville	35	49	68	83	99
Texas—Austin	51	80	125	169	218
Vermont—Burlington	15	22	35	47	58
Virginia—Roanoke	23	33	48	61	73
West Virginia—Elkins	23	31	42	51	60
Wisconsin—Madison	29	42	61	77	95

The soil-erodibility factor K has been determined experimentally. For a particular soil, it is the rate of erosion per unit of erosion index from *unit* plots on that soil. A unit plot is 72.6 feet long with a uniform slope of 9 percent in continuous fallow with tillage operations up and down the slope. These conditions for the unit plot were selected because they represent the predominant slope length and gradient on which past erosion measurements in the United States have been made.

The values of K determined for twenty-two major soil types are given in Table 4.3.

TABLE 4.3. Values of Soil-Erodibility Factor K for Major Soil Types

Soil	K	Soil	K
Dunkirk silt loam	0.69	Honeoye silt loam	0.28
Keene silt loam	0.48	Cecil sandy loam	0.28
Shelby loam	0.41	Ontario loam	0.27
Lodi loam	0.39	Cecil clay loam	0.26
Fayette silt loam	0.38	Boswell fine sandy loam	0.25
Cecil sandy clay loam	0.36	Zaneis fine sandy loam	0.22
Marshall silt loam	0.33	Tifton loamy sand	0.10
Ida silt loam	0.33	Freehold loamy sand	0.08
Mansic clay loam	0.32	Bath flaggy silt loam with 2-inch	
Hagerstown silty clay loam	0.31	surface stones removed	0.05
Austin clay	0.29	Albia gravelly loam	0.03
Mexico silt loam	0.28		

SLOPE LENGTH (L)

Slope length is defined as the distance from the point of origin of overland flow to either of the following, whichever is limiting, for the major part of the area being considered: the point where the slope gradient decreases to the extent that deposition begins or the point where runoff enters a well-defined channel. The soil loss per unit area increases as the slope length increases. This results from the greater accumulation of runoff as the slope length increases.

SLOPE GRADIENT (S)

As the slope becomes steeper, the velocity of the runoff water increases and this results in an increase in the power of the runoff water to detach particles from the soil mass and to transport them from the field.

The slope-length factor L and the slope-gradient factor S in the soil-loss equation may be taken from Figure 4.2. The factor LS is the expected ratio of soil loss per unit area on a field slope to the corresponding loss from the unit plot having a 9 percent slope, 72.6 feet long.

CROPPING MANAGEMENT (C)

The factor C in the soil-loss equation is the ratio of soil loss from land cropped under specified conditions to the corresponding loss from continuously fallow and tilled land. This factor measures the combined effect of all the interrelated cover and management variables.

The loss that would occur on a particular field if it were continuously in fallow condition is computed by the four-factor product $RKLS$ in the erosion equation. Actual loss from the cropped field is usually much less than this amount. Just how much less depends on the particular combination of cover, crop sequence, and management practices. It also depends on the particular stage of growth and development of the vegetal cover at the time of the rain. The factor C adjusts the soil-loss estimate to suit these conditions.

The correspondence of periods of expected highly erosive rainfall with periods of poor or good plant cover differs among regions or locations. Therefore, the value of C for a particular cropping system will not be the same in all parts of the country. In order to derive the appropriate C values for a given locality, it is necessary to know how the erosive rainfall in that locality is likely to be distributed through the twelve months of the year. It is also necessary to know how much erosion control protection the growing plants, the prior crop residues, and the various tillage operations will provide at the time when erosive rains are likely to occur. A procedure has been developed for deriving locational values of the factor C on the basis of available weather records and research data that reflect effects of crops and management.

Variables. The effects of cropping and management variables cannot be evaluated independently because of the many interreactions involved. Almost any crop can be grown continuously, or it can be grown in any one of numerous rotations. The sequences of crops within a system can be varied. Crop residues can be removed, left on the surface, incorpo-

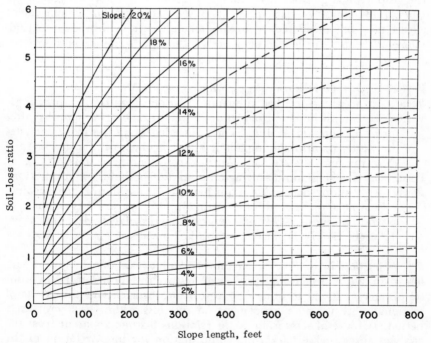

Figure 4.2. Slope-length and slope-gradient factors (LS).

rated near the surface, or plowed under. When left on the surface, they can be chopped or they can be allowed to remain as left by the harvesting operation. Seedbeds can be left rough, with much available capacity for surface storage of rainfall, or they can be left smooth. Different combinations of these variables are likely to have different effects on soil loss.

In addition, the effectiveness of crop residue management depends on how much residue there is. This in turn depends on rainfall distribution, the fertility level, and various management decisions made by the farmer. Similarly the erosion control effectiveness of meadow sod turned under before corn or other row crops depends on the type and quality of the meadow and on the length of time elapsed since the sod was turned under.

The canopy protection of crops not only depends on the type of vegetation, the stand, and the quality of growth but also varies greatly in different months or seasons. Therefore the overall erosion-reducing effectiveness of a crop depends largely on how much of the erosive rain occurs during those periods when the crop or management practice provides the least protection.

Crop-Stage Periods. The change in effectiveness of plant cover within the crop year is gradual. For computation of C values, the year is divided into crop-stage periods, defined so that cover and management effects may be considered approximately uniform within each period. (Some adjustment in length of periods 1 through 3 may be necessary for vegetable crops.)

Period F. Rough fallow—turn plowing to seeding.
Period 1. Seedling—seedbed preparation to one month after planting.
Period 2. Establishment—from end of period 1 to two months after spring or summer seeding. For fall-seeded grain, period 2 includes the winter months, ending about May 1 in the northern states, April 15 in the central states, and April 1 in the southern states.
Period 3. Growing and maturing crop—end of period 2 to crop harvest.
Period 4. Residue or stubble—crop harvest to plowing or new seeding. (When meadow is established in small grain, period 4 is assumed to extend two months beyond the grain harvest date. After that time, the vegetation is classified as established meadow.)

Effects on Soil Loss. About ten thousand plot-years of runoff and soil-loss data assembled from forty-seven research stations in twenty-four states were analyzed to obtain empirical measurements of the effects of cropping systems and management on soil loss within each crop-stage period (1). Several significant factor relations became apparent from the analyses and provide background information for interpretation of the soil-loss ratio table (Table 4.4).

Fallow soil. The rate at which fallow soil eroded depended on cropping history and the nature and quantity of residues turned under, as well as on inherent characteristics of the soil itself. Brief periods of fallow in a rotation were not comparable in erodibility to continuous clean-tilled fallow on similar soil and slope. Plant residues incorporated in fallow soil were very effective in reducing both runoff and erosion. Effects of cropping history are a part of the factor *C* in the erosion equation.

Productivity level. In general, soil losses decreased as crop yields increased. Since good grain yields are usually associated with good stands and good forage growth, the canopy cover is better and more residues

TABLE 4.4. Ratio of Soil Loss from Cropland to Corresponding Loss from Continuous Fallow

Line No.	Cover, sequence, and management [1]	Productivity [2]		Soil-loss ratio for crop-stage period [3]						
		Hay yield	Corn yield	F	1	2	3 [4]	4L	4R	4L+WC
	CORN IN ROTATION	*Tons*	*Bu.*	*Pct.*	*Pct.*	*Pct.*	*Pct.*	*Pct.*	*Pct.*	*Pct.*
	1st-year C after gr & lg hay: Spg TP, conv till									
1		3–5	75+	8	25	17	10	15	35	10
2	Do	2–3	75+	10	28	19	12	18	40	11
3	Do	2–3	60–74	12	29	23	14	20	43	13
4	Do	1–2	60–74	15	30	27	15	22	45	13
5	Do	1–2	40–59	15	32	30	19	30	50	15
6	Do	<1	40–59	23	40	38	25	35	60	18
7	Do	<1	20–35	23	40	43	30	45	65	23
8	Spg TP, pl plant	3–5	75+	------	8	8	6	15	35	10
9	Do	2–3	75+	------	10	10	7	18	40	10
10	Do	2–3	60–74	------	12	12	8	20	43	13
11	Do	1–2	60–74	------	15	15	9	22	45	13
12	Do	1–2	40–59	------	15	15	11	30	50	15
	2d-year C after gr & lg hay: RdL, spg TP, conv till									
13		3–5	75+	25	48	37	20	24	------	14
14	Do	2–3	75+	32	51	41	22	26	------	15
15	Do	2–3	60–74	35	54	45	24	28	------	15
16	Do	1–2	40–59	42	57	49	28	42	------	21
17	Do	<1	40–59	46	62	54	30	50	------	25
18	Do	<1	20–35	55	66	60	35	65	------	33
19	RdL, spg TP, pl plant	3–5	75+	------	25	25	12	24	60	14
20	Do	2–3	75+	------	32	32	13	26	60	14
21	Do	2–3	60–74	------	35	35	14	28	65	15
22	Do	1–2	40–59	------	42	42	17	42	70	21
23	Do	<1	40–59	------	46	46	18	50	75	25
24	RdL, WC in prec C, conv till	3–5	75+	18	35	30	20	24	60	14
25	Do	2–3	75+	20	37	33	22	26	60	15
26	Do	2–3	60–74	21	39	36	24	28	65	15
27	Do	1–2	40–59	25	42	40	28	42	70	21
28	Do	<1	40–59	28	45	44	30	50	75	25
29	Do	<1	20–35	33	48	49	35	65	80	33
30	RdR, WC in prec C, conv till	3–5	75+	30	52	40	22	------	60	35
31	Do	2–3	60–74	35	55	45	24	------	65	35
32	Do	1–2	40–59	42	60	53	30	------	75	35
33	RdR, no WC, conv till	3–5	75+	55	62	47	22	------	60	--------
34	Do	2–3	60–74	60	65	51	24	------	65	--------
35	Do	1–2	40–59	65	72	57	30	------	70	--------
	3d or 4th year C after gr & lg hay, or 2d-year C after SG, red cl, or sw cl: RdL, conv till									
36		3–5	75+	36	63	50	26	30	------	--------
37	Do	2–3	60–74	45	66	54	29	40	------	--------
38	Do	1–2	40–59	55	70	58	32	50	------	--------
39	Do	<1	20–35	70	76	64	38	65	------	--------
40	RdL, pl plant	3–5	75+	------	36	36	16	30	------	--------
41	Do	2–3	60–74	------	45	45	17	40	------	--------
42	Do	1–2	40–59	------	55	55	19	50	------	--------
43	RdL + WC in prec C	3–5	75+	22	46	41	26	30	------	15
44	Do	2–3	60–74	26	48	44	29	40	------	20
45	Do	1–2	40–59	33	51	47	32	50	------	25

TABLE 4.4. *(Continued)*

Line No.	Cover, sequence, and management [1]	Productivity [2] Hay yield	Corn yield	Soil-loss ratio for crop-stage period [3] F	1	2	3 [4]	4L	4R	4L+WC
		Tons	*Bu.*	*Pct.*	*Pct.*	*Pct.*	*Pct.*	*Pct.*	*Pct.*	*Pct.*
	CORN IN ROTATION—continued									
46	RdL+WC in prec C	<1	20–35	42	56	52	38	65	------	33
47	RdR, conv till	3–5	75+	70	78	54	27	------	62	--------
48	Do	2–3	60–74	75	80	60	30	------	70	--------
49	Do	1–2	40–59	75	80	70	35	------	75	--------
50	RdR, 8 tons manure added	--------	60–74	60	70	52	28	------	62	--------
51	1st-year C after cl hay	2	40–55	21	35	32	25	35	60	--------
52	1st-year C after sw cl	---	40–55	23	45	38	28	35	60	--------
53	1st-year C after lesp hay	1–2	60–70	55	70	55	30	40	65	--------
54	Do	1–2	40–55	55	70	60	32	50	75	--------
55	C after 1 year cot after gr & lg hay	2–3	60–70	30	58	46	24	28	65	--------
56	Do	1–2	40–59	35	65	54	29	42	70	--------
	Corn in meadowless systems:									
57	After SG wintercrop, spg TP	--------	75+	22	37	35	22	27	------	--------
58	Do	--------	60–70	25	40	38	24	30	------	--------
59	Do	--------	40–55	30	45	42	30	40	------	--------
60	After SG, no intercrop, RdL	--------	--------	(5)	(5)	(5)	(5)	(5)	------	--------
	Corn after corn (lines 129–34)									
	COTTON IN ROTATION									
61	1st-year cot after gr & lg hay	3–5	HP	8	25	30	20	22	------	15
62	Do	2–3	HP	10	30	35	25	25	------	16
63	Do	1–2	HP	15	34	40	30	30	------	18
64	Do	1–2	MP	15	34	45	35	33	------	20
65	Do	<1	MP	23	40	54	45	42	------	23
	2d-year cot after gr & lg hay:									
66	RdL, no WC seeding	3–5	HP	30	54	56	38	38	------	20
67	Do	2–3	HP	34	58	62	44	40	------	20
68	Do	1–2	MP	40	65	68	46	42	------	22
69	Do	<1	MP	45	70	70	50	48	------	25
70	RdL+WC in prec cot	3–5	HP	20	40	46	38	38	------	20
71	Do	2–3	HP	23	42	50	44	40	------	20
72	Do	1–2	MP	23	47	55	46	42	------	22
73	Do	<1	MP	27	51	57	50	48	------	25
	Cot after cot, 3d or more year after M:									
74	RdL, no WC seeding	--------	HP	42	70	70	48	42	------	22
75	Do	--------	MP	45	80	80	52	48	------	25
76	RdL+WC in prec cot	--------	HP	32	51	57	48	42	------	22
77	Do	--------	MP	35	58	65	52	48	------	25
78	Cot after 1 year C (RdL) after M	3–5	75+	25	48	49	32	38	------	20
79	Do	2–3	60–75	32	51	51	35	40	------	20
80	Do	1–2	40–59	35	54	56	38	45	------	23
81	Cot after 1 year C, C RdR	2–3	40–59	60	65	63	40	48	------	25
82	Cot after 2 years C (RdL) after M	3–5	75+	36	63	62	39	45	------	23
83	Do	2–3	60–75	45	66	68	45	48	------	25
84	Do	1–2	40–59	55	70	73	50	48	------	25
85	Cot in cot (V)–C (crot) system	--------	HP	28	40	45	35	------	------	22
86	Cot in cot-O-lesp seed, RdL	--------	HP	23	34	40	30	------	------	--------
87	Do	--------	MP	25	40	45	37	------	------	--------
88	Cot in cot-SG-sw cl	--------	MP	25	45	48	35	------	------	--------
	SMALL GRAIN IN ROTATION									
	With meadow seeding: In disked row-crop residue—									
89	After 1 year C after M	3–5	75+	------	20	12	2	2	------	--------
90	Do	2–3	60–74	------	30	18	3	2	------	--------
91	Do	1–2	40–59	------	41	25	4–15	2	------	--------
92	Do	<1	25–39	------	60	36	5–15	3	------	--------
93	After 2d or 3d year C after M	3–5	75+	------	32	19	5	3	------	--------
94	Do	2–3	60–74	------	40	24	5	3	------	--------
95	Do	1–2	40–59	------	58	35	5–15	3	------	--------
96	Do	<1	25–39	------	75	45	6–15	3	------	--------
97	After 1 or more C after SG	--------	--------	------	(6)	(6)	(6)	(6)	------	--------

TABLE 4.4. *(Continued)*

Line No.	Cover, sequence, and management [1]	Productivity [2]		Soil-loss ratio for crop-stage period [3]						
		Hay yield	Corn yield	F	1	2	3 [4]	4L	4R	4L+WC
	SMALL GRAIN IN ROTATION—continued									
	With meadow seeding—Continued									
	In disked row-crop residue—Con.	*Tons*	*Bu.*	*Pct.*	*Pct.*	*Pct.*	*Pct.*	*Pct.*	*Pct.*	*Pct.*
98	After 1st-year cot after M	2–3			35	25	5–15	3		
99	After 2d-year cot after M	2–3			50	35	5–15	3		
100	In cot middles after sw cl or lesp				30	22	10–15	3		
	On disked row-crop stubble, RdR—									
101	After 1 year C after M	2–3	60+		50	40	5–15	3		
102	Do	1–2	40–59		80	45	7–15	3		
103	After 2 years C after M	2–3	60+		80	50	6–15	3		
104	After C, 3d year after M				92	55	7–15	3		
	On plowed seedbed, RdL—									
105	After 1 year C or SG after M	3–5	75+	25	45	30	5	3		
106	Do	2–5	60–74	35	51	34	5	3		
107	Do	1–2	40–59	42	60	40	7	4		
108	After 2 years C or SG after M	3–5	75+	36	60	40	5	3		
109	Do	2–3	40–59	55	70	45	7	4		
	On plowed seedbed, RdR—									
110	After 1 year C or SG after M	3–5	75+	55	60	40	5	3		
111	Do	2–3	60–74	60	65	42	6	3		
112	Do	1–2	40–59	65	70	45	7	4		
113	After 2 years C after M	2–3	60–74	65	70	45	7	4		
	Without meadow seeding:									
114	Sequences and yields of lines 89–90			(⁷)	(⁷)	(⁷)	8	8	16	
115	Sequences and yields of lines 91–99, 101, 105, 106, 108–110			(⁷)	(⁷)	(⁷)	10	10	20	
116	Sequences and yields of lines 102–104, 107, 111–113			(⁷)	(⁷)	(⁷)	12	12	25	
	DOUBLE-CROPPED ROTATIONS									
117	Wheat (grain) and lesp (hay)				25	25	5	5		
118	Wheat and lesp, both grazed				25	25	12	6		
119	Spg oats (hay) and lesp (hay) [8]				50	18	5	5		
	ESTABLISHED MEADOWS [9]									
120	Grass and legume mix	3+						0.4		
121	Do	2						.6		
122	Do	1						1.0		
123	Alfalfa	2.5+						2.0		
124	Lespedeza							2.0		
125	Red clover							1.5		
126	Sericea, 2d year							2.0		
127	Sericea, after 2d year							1.0		
128	Sweetclover							2.5		
	CORN AFTER CORN									
	Strip till. Planted with indicated amount of residue on the surface.									
129	1,000 to 1,500 lbs		75+		50	40	25	30		
130	1,500 to 2,000 lbs		75+		40	32	18	25		
131	2,000 to 3,000 lbs		75+		30	24	14	20		
132	3,000 to 4,000 lbs		75+		20	16	9	15		
133	4,000 to 6,000 lbs		75+		7	7	7	7		
134	> 6,000 lbs		75+		3	3	3	3		

[1] Symbols: C, corn; conv till, conventional tillage; pl plant, plow plant; cot, cotton; crot, crotolaria; gr & lg, grass and legume; lesp, lespedeza; M, grass and legume meadow, at least 1 full year; O, oats; prec, preceding; RdL, residue of prior crop left; RdR, residue of prior crop removed; spg, spring; SG, small grain; sw cl, sweetclover; TP, turn plow; V, vetch; WC, grass or grass-and-legume winter cover seeded early.

[2] For cotton, HP = high crop productivity; MP = moderate crop productivity.

Small-grain cover is assumed commensurate with the indicated productivity level of corn or cotton.

[3] Period 4 ratios are taken from column 4L when crop residues remain on field but without winter cover seeding; from column 4R when corn stover, straw, and similar residues are removed; and from column 4L + WC when early-seeded grass and legume winter cover is established in addition to leaving crop residues.

[4] Where two period 3 values appear, the first is for high-yielding grain and the second is for grain yielding less than 30 bushels of oats or 15 bushels of wheat per acre.

[5] Use data from lines 36 to 42, selecting line on basis of productivity level.

[6] Use data from lines 93 to 96, selecting line on basis of productivity level.

[7] Use data from lines 89 to 113.

[8] Ratio for winter months is 12 percent.

[9] Ratios shown are the yearly averages.

49

are returned to the soil. Both help to decrease erosion losses. However, the added erosion-reducing benefit of each additional unit of crop yield becomes less as yields become higher.

Crop-residue. The soil-loss reduction resulting from prior crop residues left on the field depended on the type and quantity of residues produced and the method of handling. Residues were most effective when left at the surface, but after several years of turning heavy crop residues under with a moldboard plow before row-crop seeding, both runoff and soil loss from the row crop were much less than from similar plots from which cornstalks and grain straw were removed at harvest time. The effectiveness of incorporated residues was greatest during the fallow and seedling periods.

Row crops after meadow. Specific year erosion losses from corn after meadow ranged from 14 to 68 percent of corresponding losses from continuous corn on adjacent plots. Mixtures of grass and legume were more effective than legumes alone. In general, the effectiveness of grass and legume meadow sod plowed under before corn in reducing soil loss from the corn was directly proportional to meadow yields. Its erosion control effectiveness was greatest during the rough fallow and corn seedling periods and decreased thereafter. The total reduction in soil loss effected by the meadow depended therefore largely upon the stage of development of the corn when the erosive rains occurred. The length of the period during which the turned sod remained effective in reducing erosion was also directly related to meadow yields.

Length of meadow periods. Direct comparisons of corn after first, second, and third years of meadow were very limited, and the data were too sporadic for overall differences to be statistically significant. When second year meadow was allowed to deteriorate under poor management, it was less effective than one year of meadow. When succeeding meadows were more productive than the first year, they were usually more effective in reducing erosion from corn after the meadow. The effectiveness of virgin sod and of long periods of continuous alfalfa in which grass became well established was longer lasting than that of one or two years of rotation meadow.

Grass and legume catch crops, established in spring-seeded small grain and plowed under at corn planting time in the following year, effected significant reduction in soil erosion during the corn seedling period, but their effectiveness was shorter lived than that of a full year of meadow.

Winter-cover seedings. The erosion control attained with winter-cover seedings depended upon time and method of seeding, time of plowing, rainfall distribution, and type of cover seeded. Covers such as vetch and ryegrass, seeded between the corn or cotton rows before harvest and turned under in April, were effective in reducing erosion not only in the winter months but also during the seedling and establishment periods of the following crop. Small grain alone, seeded in corn or cotton residues and plowed under in spring, was of some value during the winter peri-

TABLE 4.5. Suggestions for Approximating Soil-Loss Ratios for Cropping and Management Combinations Not Listed in Table 4.4

Cover, sequence, and management	Soil loss ratios
Corn:	
After fall turnplowing in northern half of United States.	To compensate for effect of freezing and thawing and for high early-spring soil moisture content, add 7 to each period-F and period-1 value in lines 1 to 7, 13 to 18, 33 to 39, and 47 to 50 of table 4.4.
After 2 or more full years of meadow.	Table 4.4 assumed at least 1 full year of established grass-and-legume meadow. Additional credit for 2-year meadows may be considered if meadows are high yielding and are not permitted to deteriorate: Reduce by 10 percent the values for periods F, 1, 2, 3 and 4L in lines 13, 14, 15, 33, 34, 66, 67, 78, and 79.
With small-grain seeding for winter cover.	Small grain turned early in spring does not significantly reduce soil loss from following corn crop. Select lines from table 4.4 that do not specify WC seeding and substitute small-grain periods 1 and 2 for corn period 4L or 4R.
Grain sorghum_____	Same as ratios for corn in similar rotations where canopy cover and quantities of residue are comparable. Under irrigation, the values for grain sorghum may equal those for high-yielding corn.
Meadow:	
New_____	When seeded without a nurse crop, use values listed for spring-seeded small grain. The lengths of periods 1 and 2 should be adjusted if necessary so that cover in each period will be comparable to corresponding grain periods.
Established_____	Apply values of lines 120 to 128.
Peanuts_____	For comparable crop sequence, values in lines 5, 6, 16, 17, 27, 28, 32, 35, 38, 45, 49, 52, and 56 are recommended.
Potatoes:	
After potatoes or truck crop_____	In similar crop sequence, select values from periods F, 1, 2, and 4 of lines 18, 29, 39, 46. For period 3, use values from line 16, 17, 27, 28, 38, or 45.
After grass-and-legume hay yielding more than 2 tons per acre.	Select values for periods F to 3 from lines 1, 3, 5, 7 on basis of hay yield; period 4 from line 7.
After corn or small grain_____	Select values for periods F to 3 on basis of preceding crop and yield; period 4 from line 7.
Soybeans:	
After grass-and-legume hay or after corn.	Use values for comparable corn rotations: Periods F to 3 from lines 3 to 7, 15 to 18, 26 to 29, 37 to 39, 48, 49; period 4 from lines 7, 18, 29, 39.
After soybeans_____	Select lines representing corn residues equivalent to soybean residues: Lines 15 to 18, 37 to 39.
Late-planted_____	Select values from comparable crop sequences in lines 5, 6, 16, 17, 27, 28, 32, 35, 38, 45, 49.
Sweet corn_____	Do.
Truck crops_____	For low-residue truck crops after grass-and-legume hay or high-residue crops, select periods F and 1 values from comparable corn rotations; periods 2, 3, and 4 values from lines 7, 18, 29, 39, 46. For second or more year of truck crop, use values from line 39 or 46 for all periods.

od but showed no residual erosion-reducing effect after the next year's corn or cotton planting. Very limited data indicated crimson clover alone to be of doubtful value as a winter cover, but when it was combined with a quick starting grass, effective protection was provided. Cotton land that was plowed in the fall and seeded to small grain or vetch which was plowed under in spring for another cotton seeding lost about 20 percent more soil than adjacent land with undisturbed cotton residues on the surface.

Type of tillage. Conventional tillage consists of plowing with either a disk or moldboard plow, plus one or more disk harrowings, followed by a spring or spike-tooth harrow. Plow planting consists of seedbeds prepared by plowing followed immediately by planting. Strip tillage consists of tilling a narrow strip in the residue of the previous crop.

Climate. The soil-loss ratio varies depending on the climate of the area.

Humid areas. An empirical measure of the erosion control effectiveness of each crop grown in various sequences in humid areas was obtained from assembled Agricultural Research Service plot data (1). Ratios of soil losses from the cropped plots to corresponding losses from continuous fallow were computed. This ratio was computed for each of the five crop-stage periods and for each particular crop in various combinations of sequence and productivity level.

Table 4.4 lists, for each crop-stage period, the expected ratio of soil loss from the designated crop and practice combination to corresponding loss from the base fallow condition. The table is entered on the basis of crop, crop sequence, residue management, and crop productivity level, in that order. The soil-loss ratio for each crop-stage period is taken from the seven columns at the right. Three columns are needed for period 4, in order to reflect effects of different ways of managing the crop. Suggestions for estimating soil-loss ratios under some of the conditions not directly listed in this table are shown in Table 4.5.

Semiarid areas. Water erosion is also a serious problem in most semiarid regions. Inadequate moisture and periodic droughts reduce the periods when growing plants provide good soil cover and limit the total quantities of plant residues produced. Erosive rainstorms are not uncommon, and they are concentrated within the season when cropland is least protected. Because of the difficulty of establishing meadows and the competition for available soil moisture, sod-based rotations are often impractical.

Proper management of available residues offers one of the most important opportunities for a higher level of soil and moisture conservation. However, accurate soil-loss ratios for stubble mulching and summer fallowing practices in the western Great Plains area are not yet available from research data. The ratios given in Table 4.6 are approximations, based on observations of experienced field personnel guided by very limited data on the erosion control effectiveness of various amounts of surface mulch and by the experimentally determined values of Table 4.4. These approximations appear to be consistent with present knowledge of erosion research and runoff and provide valuable guides until more precise evaluations can be obtained through additional research.

Erosion Index Distribution. The rainfall factor R in the erosion equation does not completely describe the effects of locational differences in rainfall pattern on soil erosion. The erosion control effectiveness of a cropping system on some particular field depends in part on how the year's erosive rainfall is distributed among the five crop-stage periods of each crop included. Therefore expected distribution of erosive rainfall at a particular location is an important element in deriving the applicable value of the cropping-management factor C.

On the basis of the distribution of the erosion index throughout the year, the thirty-seven states of Figure 4.1 were divided into thirty-three geographic areas shown in Figure 4.3. The average erosion index

TABLE 4.6. Approximations of Soil-Loss Ratios for Crop-Stage Periods and Number of Tillage Operations with Stubble Mulching and Summer Fallowing in Western Great Plains Areas

Cover, sequence, and management	Residue on surface at seeding time	Soil-loss ratios for crop-stage period—				
		1	2	3	4L	4R
Small grain without meadow seeding:	*Pounds*	*Pct.*	*Pct.*	*Pct.*	*Pct.*	*Pct.*
After small grain	200–500	70	45	6	10	20
Do	500–1,000	42	25	6	10	20
Do	1,000–1,500	25	17	6	10	20
Do	1,500–2,000	15	10	6	10	20
After summer fallow of—						
Small-grain residues	0–200	90	55	6	10	20
Do	200–500	70	45	6	10	20
Do	500–1,000	42	25	6	10	20
Do	1,000–1,500	25	17	6	10	20
Do	1,500–2,000	15	10	6	10	20
Row crop residues	0–200	90	55	6	10	20
Do	200–500	85	50	6	10	20
Do	500–1,000	70	45	6	10	20
Do	1,000–1,500	50	35	6	10	20
Do	1,500–2,000	40	30	6	10	20
Do	2,000–2,500	30	25	6	10	20

		Soil-loss ratios for following number of tillage operations after grain harvest—				
		1	2	3	4	5
Summer fallow:						
After small grain	0–200	53	60	70	80	90
Do	200–500	25	49	55	63	70
Do	500–1,000	25	29	34	39	42
Do	1,000–1,500	10	14	19	22	25
Do	1,500–2,000	4	6	8	11	13
After row crop	0–200	68	72	80	85	90
Do	200–500	50	55	63	75	85
Do	500–1,000	50	55	60	65	70
Do	1,000–1,500	26	35	40	45	50
Do	1,500–2,000	20	25	30	35	40
Do	2,000–2,500	15	20	25	28	30

value at a given date expressed as a percentage of the average annual value for each of the thirty-three subareas of Figure 4.3 is shown in Table 4.7. The percentage of the annual erosion index that is to be expected within any particular crop-stage period may be found by subtracting the value for the first date from the value for the last date.

Deriving a C Value for a Specific Locality. To compute the value of C for any particular rotation on a given field, it is necessary first to determine the most likely seeding and harvest dates, method of seedbed preparation and residue management, and average crop yields to be expected with this system on the soil involved and with the contemplated management. The procedure to follow in evaluating C will be explained by means of an example.

Example: Evaluate C for a four-year rotation of wheat with meadow seeding, meadow, corn, and corn in central Indiana (area 16) with conventional tillage and average production of 45 bushels of wheat, 4 tons of hay, and 100 bushels of corn per acre. Assume that the meadow is a

TABLE 4.7. Percentage of the Average Annual Erosion Index Value Which Will Occur on a Given Date after January 1. Computed for the Geographic Areas Given in Figure 4.3

Date		1	2	3	4	5	6	7	8	9	10	11	12	13	14	15	16	17	18	19	20	21	22	23	24	25	26	27	28	29	30	31	32	33
Jan.	1	0	0	0	0	0	0	0	0	0	0	0	0	0	0	0	0	0	0	0	0	0	0	0	0	0	0	0	0	0	0	0	0	0
	10	0	0	0	0	1	0	0	0	1	0	1	0	0	0	0	0	0	1	1	1	3	2	2	2	1	1	0	1	0	0	0	0	1
	20	0	0	0	0	1	0	0	2	2	1	2	0	0	0	0	0	1	2	2	2	5	4	3	4	2	3	1	3	0	1	0	1	2
Feb.	1	0	0	0	0	2	0	1	3	4	2	3	0	0	0	0	1	1	3	3	3	7	7	5	7	3	4	2	4	1	1	1	1	3
	10	0	0	0	1	3	0	2	4	5	3	4	0	0	1	1	2	2	4	5	4	9	9	7	8	4	5	3	5	2	2	1	2	3
	20	0	0	0	1	4	0	3	5	7	4	6	0	0	1	1	3	3	5	7	6	12	11	9	11	6	7	4	6	3	3	1	3	4
March	1	0	1	0	2	4	0	3	7	9	6	7	0	1	2	2	4	3	6	9	7	14	13	11	13	7	9	5	7	4	3	2	3	6
	10	0	1	1	3	5	0	4	9	12	8	8	1	2	3	3	5	4	8	12	9	15	16	13	15	9	12	7	9	5	4	3	4	8
	20	0	2	1	4	7	0	5	11	14	10	9	1	3	4	4	6	5	9	14	12	17	19	16	18	11	14	8	10	6	5	3	5	9
April	1	0	3	2	5	10	1	6	14	17	12	10	2	3	5	5	8	6	11	17	14	19	22	18	20	13	16	10	12	7	6	4	6	11
	10	1	4	3	7	13	2	8	18	21	15	13	3	4	6	7	10	7	13	19	15	21	25	21	23	15	19	12	14	9	7	5	7	13
	20	2	5	4	9	17	4	10	22	25	18	15	4	5	8	9	12	8	16	21	17	24	28	24	26	17	22	15	16	10	8	6	8	14
May	1	3	7	6	12	21	8	14	28	29	22	18	5	7	10	11	14	11	20	26	19	26	33	27	29	19	25	18	17	11	10	7	10	15
	10	5	9	9	16	26	13	22	34	33	27	23	8	10	12	14	17	13	25	30	21	28	37	30	32	22	28	21	20	13	13	10	12	17
	20	7	13	15	21	32	20	32	41	38	33	29	11	14	16	18	21	16	30	33	24	31	41	33	34	25	33	24	25	14	16	14	14	18
June	1	11	19	23	27	37	27	40	48	42	39	34	16	19	20	23	24	20	35	37	27	34	44	35	37	28	36	26	25	17	19	17	17	21
	10	18	25	33	34	43	34	47	53	46	44	38	23	28	25	28	30	25	40	41	31	37	48	38	41	32	40	29	28	20	23	21	20	24
	20	27	33	43	41	49	40	53	57	50	49	42	30	40	31	34	39	32	45	46	36	41	52	41	45	36	44	33	32	24	29	24	25	28
July	1	36	42	51	47	54	45	57	60	54	53	46	38	47	38	40	44	39	50	50	41	47	55	45	51	40	48	37	36	30	35	33	32	32
	10	45	51	58	52	58	50	60	63	57	55	49	46	54	47	47	53	47	54	54	47	54	59	49	56	44	54	42	41	38	40	39	39	37
	20	54	58	64	57	61	55	63	65	59	58	53	54	59	55	53	59	55	58	59	55	61	63	54	62	48	61	49	48	46	47	47	45	41
Aug.	1	63	67	70	62	64	60	67	67	62	61	57	62	65	62	60	64	63	63	64	62	68	67	60	68	53	68	57	56	54	55	55	51	46
	10	72	74	75	66	67	65	70	70	65	64	60	70	71	68	67	68	70	67	68	68	73	70	65	73	58	73	66	63	61	63	63	57	52
	20	81	80	80	70	70	70	72	73	67	67	64	78	76	74	73	74	76	71	73	74	77	72	70	77	63	78	74	70	68	69	70	63	58
Sept.	1	90	85	84	74	74	74	76	77	70	70	68	84	81	79	79	78	81	74	77	79	80	75	74	80	69	81	80	76	75	75	75	68	64
	10	93	89	88	80	78	79	78	80	72	74	71	88	86	83	83	83	85	77	79	83	82	77	78	82	75	84	85	81	81	80	81	73	69
	20	96	93	91	85	83	84	81	83	75	77	75	91	90	87	87	85	88	81	82	86	84	79	81	84	82	86	89	84	85	83	85	77	73
Oct.	1	98	95	93	89	87	89	85	85	78	82	79	94	93	90	91	88	91	84	84	88	86	81	83	87	88	88	92	87	89	86	89	81	77
	10	99	97	95	93	90	94	89	88	80	86	83	96	95	93	93	91	93	87	86	90	88	83	85	88	90	89	93	89	92	88	91	84	80
	20	100	98	96	95	93	97	95	90	83	88	86	97	97	95	94	93	94	90	88	91	89	84	87	90	93	90	94	91	94	90	93	87	82
Nov.	1	100	99	97	97	95	98	98	92	85	91	89	98	98	96	95	95	95	92	90	92	90	86	88	92	94	91	95	93	95	93	94	89	85
	10	100	100	98	98	96	99	99	94	88	93	92	98	98	97	96	96	96	94	92	94	92	88	90	93	95	93	95	94	96	95	95	91	88
	20	100	100	99	99	97	100	99	96	90	95	94	99	99	98	97	97	97	96	94	95	93	91	91	95	95	94	96	96	97	96	97	94	91
Dec.	1	100	100	100	100	98	100	99	98	93	97	96	99	99	99	98	98	98	97	95	96	94	94	93	97	96	95	97	97	98	97	98	96	94
	10	100	100	100	100	99	100	100	99	95	98	98	100	100	99	99	99	99	98	95	97	96	98	95	98	97	97	98	98	99	98	99	97	96
	20	100	100	100	100	100	100	100	100	98	99	99	100	100	100	100	100	100	98	98	98	98	98	98	99	98	98	99	99	100	99	100	98	98

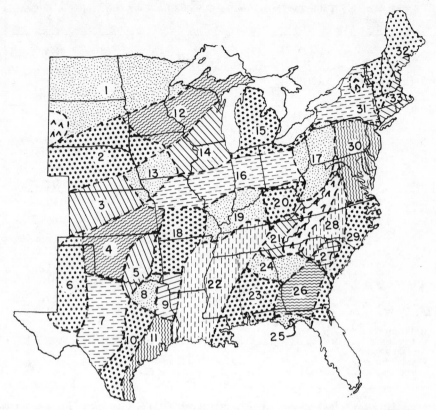

Figure 4.3. Geographic areas based on distribution of the erosion index.

mixture of grass and legume such as alfalfa and brome or timothy and clover, that crop residues are left on the field, that cornstalks are plowed under about May 1 for corn planting or disked for wheat seeding about October 10, and that wheat is harvested about July 10.

Procedure: Set up a working table such as that illustrated in Table 4.8, obtaining the needed information as follows:

Column 1 lists in chronological sequence all plowing, seeding, and harvest operations (other than hay) involved in the rotation.

Column 2 lists the beginning date of each successive crop-stage period. A seeding date begins crop-stage period 1. By definition, period 2 begins one month later, period 3 begins two months after seeding (except for winter grain), period 4 begins with crop harvest, and period F with the date of moldboard plowing. The meadow period begins two months after wheat harvest and extends to plowing date. Thus all the dates in column 2 are determined by the locational seeding and harvest dates.

TABLE 4.8. Working Table for Derivation of C Value for Four-Year Rotation in Central Indiana

(1) Operation	(2) Date	(3) Values from Table 4.7	(4) Crop-stage period	(5) Erosion index in period	(6) Soil-loss ratio [1][2]	(7) Column 5 times col. 6	(8) Value of C
		(%)		(%)	(%)		
Pl W_____	10/10	91	_____	_____	_____	_____	
	11/10	96	W1	5	32(93)	0. 0160	
	5/1	114	W2	18	19	. 0342	
Hv W_____	7/10	153	W3	39	5	. 0195	
	9/10	183	W4	30	3	. 0090	0. 0787
	9/10	283	M	100	. 4(120)	. 0040	
TP_____	5/1	314	M	31	. 4	. 0012	. 0052
Pl C_____	5/20	321	F	7	8(1)	. 0056	
	6/20	339	C1	18	25	. 0450	
	7/20	359	C2	20	17	. 0340	
Hv C_____	10/15	392	C3	33	10	. 0330	. 1176
TP_____	5/1	414	C4	22	15	. 0330	. 0330
Pl C_____	5/20	421	F	7	25(13)	. 0175	
	6/20	439	C1	18	48	. 0864	
	7/20	459	C2	20	37	. 0740	
Hv C & Pl W_____	10/10	491	C3	32	20	. 0640	. 2419
Rotation total, 4 years_____	_____	_____	_____	400	_____	_____	. 4764
Annual average C value for rotation_____	_____	_____	_____	_____	_____	_____	. 119

Symbols: Pl W, plant wheat; Hv W, harvest wheat; TP, turn plow; Pl C, plant corn; Hv C, harvest corn.
[1] For 45 bu. wheat, 4 tons hay, 100 bu. corn per acre, conventional seedbed and tillage.
[2] Numbers in parentheses refer to line numbers in table 4.4.

Column 3 lists the percentage of the annual erosion index for central Indiana (area 16), which is obtained from Table 4.7. A value is read for each successive date listed in column 2, adding one to the "hundreds" column each time January 1 is passed.

Column 4 identifies the crop-stage period ending with the date shown on that line.

Column 5 lists the percentage of the erosion index applicable to each successive crop-stage period. The values are differences between successive values recorded in column 3.

Column 6 lists the soil-loss ratio indicated in Table 4.4 for the specific conditions and crop-stage period represented by each line in the working table. The numbers in parentheses indicate the lines in Table 4.4 from which the values were taken. (The crop-yield figure for entering Table 4.4 is the expected average yield, not the yield attainable in the most favorable years. If the likelihood of meadow failure is significant, a yield figure well below the expected average is appropriate. From an erosion viewpoint, the adverse effects of a meadow failure in a rotation far outweigh the gains from occasional exceptionally good meadows. All row-crop values in the table that are not otherwise identified assume moldboard plowing, smoothing for seedbed, and cultivation after emergence. The F period precedes the crop year with which it is associated in the table. For example, the value for rough fallow after first year corn appears in the line for second year corn.)

Column 7 is self-explanatory. The decimals in this column derive from the percentage values in columns 5 and 6.

Column 8 subtotals for the different crops indicate where in the rotation most of the erosion is occurring and help to suggest where additional conservation measures could be most helpful in reducing it. The total for this column, divided by the number of years in the rotation, is the C value for this rotation under the conditions assumed in columns 1 to 6.

The C values computed for the rotation in the example are directly applicable only within area 16. For other areas or other seeding dates, the values in columns 3 and 5 of Table 4.8 are different and the C value for the same rotation in other areas may be either larger or smaller.

All these detailed computations are not required for each farm field. The procedure and basic data for derivation of the C values are provided primarily to enable computation of ready reference handbook tables of values applicable in specific states or geographic areas, such as Table 4.9 which has been prepared for area 16. Knowledge of the procedure will however lead to a better understanding of the significance of such tables and will permit field computation of values for unusual situations.

EROSION CONTROL PRACTICE (P)

In general, whenever a sloping field is to be cultivated and exposed to erosive rains, the protection offered by sod or close growing crops in the system needs to be supported by practices that slow the runoff water and thus reduce the amount of soil it can carry. The most important of these supporting practices for cropland are contour tillage, contour strip-cropping, and terrace systems. The factor P in the erosion equation is the ratio of soil loss with the supporting practice to the soil loss with uphill and downhill culture. Improved tillage practices, sod-based rotations, fertility treatments, and greater quantities of crop residues left on the field contribute materially to erosion control. However, these are considered conservation cropping and management practices, and the benefits derived from them are included in the factor C.

Contour Tillage. The practice of tillage and planting on the contour has been, in general, effective in reducing erosion. In limited field

TABLE 4.9. Partial List of C Values for Common Rotations in Erosion Index Distribution Area 16

Cropping System	Residue Removed		Residue Left over Winter			Plow plant
	Fall plow		Spring plow			
	Corn yield		Corn yield			Corn yield
	60–74	75+	40–59	60–74	75+	75+
C (continuous)[a]	.61	.57	.48	.43	.38	.27
C-C-C-O$_x$	.49	.45	.36	.31	.27	.19
C-C-O$_x$	.45	.41	.32	.27	.24	.17
C-O$_x$	.37	.33	.25	.20	.17	.12
C-C-C-O-M	.31	.28	.25	.20	.17	.12
C-C-O-M	.23	.21	.19	.14	.12	.082
C-C-O-M-M	.18	.17	.15	.12	.10	.066
C-C-O-M-M-M	.15	.14	.13	.096	.082	.057
C-O-M	.13	.11	.12	.079	.058	.037
C-O-M-M	.10	.084	.088	.060	.045	.029
C-O-M-M-M	.082	.068	.071	.049	.036	.024
C-O-M-M-M-M	.069	.058	.060	.042	.031	.020
M (continuous)	...	...	...	.006	...	...

Symbols: C = corn, O = oats, O$_x$ = oats with sweet clover intercrop, M = meadow.
[a] Continuous corn, strip till, 75+ yield, planted with indicated amount of residue on the surface: 2,000 lb, C value 0.20; 3,000 lb, C value 0.14; 4,000 lb, C value 0.07; 6,000 lb, C value 0.03.

studies, the practice provided almost complete protection against erosion from individual storms of moderate to low intensity, but it provided little or no protection against the occasional severe storms that caused extensive breakovers of the contoured rows. Contouring appears to produce its maximum average effect on slopes in the 3 to 7 percent range. As land slope decreases, it approaches equality with the contour row slope, and the soil-loss ratio approaches 1.0. As slope increases, contour row capacity decreases, and the soil-loss ratio again approaches 1.0.

After available data and observations were considered, the values of P shown in Table 4.10 were adopted. These are average values for the factor. Location values may vary with soil type, cropping, residue management, and rainfall pattern.

The full benefits of contouring are obtained only on fields relatively free from gullies and depressions other than grassed waterways. The effectiveness of this practice is reduced if a field contains numerous small gullies and rills that are not obliterated by normal tillage operations. In such instances, land smoothing should be considered before contouring.

Contour listing, with the corn planted in the furrows, has been more effective than surface planting on the contour. However, the additional effectiveness of this practice is limited to the time from the date of listing to that of the second corn cultivation. The soil-loss ratios (Table 4.4) that apply to this period may be reduced 50 percent in addition to

TABLE 4.10. Practice Factor Values for Contouring

Land Slope (%)	P Value
1.1–2	0.60
2.1–7	0.50
7.1–12	0.60
12.1–18	0.80
18.1–24	0.90

the reduction supplied by the contour factor. The additional credit does not apply after the lister ridges have been largely obliterated by two cultivations.

When rainfall exceeds infiltration and surface detention in large storms, breakovers of contour rows often result in concentrations of runoff that tend to become progressively greater with increases in slope length. Therefore on slopes exceeding some critical length, the amount of soil moved from a contoured field may approach or exceed that from a field on which each row carries its own runoff water down the slope. The slope length at which this would be expected to occur would depend to some extent on gradient, soil properties, management, and storm characteristics. Terraces or the sod strips in a contour strip-crop system function to prevent serious erosion damage when excessive row breakage occurs.

The values shown in Table 4.11 are guides to slope-length limits for effective contouring. These are judgment values. Research data are not available to verify or correct them. It is important to bear in mind however that the contour P values given in Table 4.10 assume slope lengths short enough for full effectiveness of the practice. On longer slopes the contour factor values are valid only for the upper portion of the slope, within the slope-length limits given in Table 4.11.

Contour Strip-cropping. Strip-cropping, a practice in which contour strips of sod are alternated with strips of row crop or small grain, has proved to be a more effective practice than contouring alone. A good example is found in the Mormon Coulee near La Crosse, Wisconsin, where some fields are reported to have been cropped in strips for more

TABLE 4.11. Slope-Length Limits for Contouring

Land Slope (%)	Maximum Slope Length (ft)
2	400
4–6	300
8	200
10	100
12	80
14–24	60

than seventy years. Where the strips were on the contour, or nearly so, good erosion control was accomplished. Where the strips were 5 percent or more off contour, very high soil losses have occurred as a result of the flow of runoff down the rows at high velocities (1).

Observations from strip-crop studies have indicated that much of the soil washed from a cultivated strip was filtered out of the runoff as it spread within the first several feet of the adjacent sod strip. Thus the strip-crop factor, derived from soil-loss measurements at the foot of the slope, accounts for off-the-field movement of soil, but not for all movement within the field. Recommended widths for strip-cropping are shown in Table 4.12.

If the strip guidelines are level and a grass strip alternates with a grain strip, the P factor should be one-half of that for contouring alone given in Table 4.10. With less effective strip-crop systems, larger factor values are recommended.

With a cropping system such as a four-year system of small grain, meadow, and two years of row crop, the P factor value should be about 75 percent of the value in Table 4.10 for contouring alone. Alternate strips of fall-seeded grain and row crop were effective on relatively flat slopes, but alternate strips of spring-seeded grain and corn on moderate to steep slopes have not appeared to provide significant erosion control benefits beyond those attained with contouring alone. For such systems the contour values are recommended.

Buffer strip-cropping consists of narrow protective strips alternated with wide cultivated strips. The location of the protective strips is determined largely by the width and arrangement of adjoining strips to be cropped in the rotation and by the location of steep, severely eroded areas on slopes. Buffer strips usually occupy the correction areas on sloping land and are seeded to perennial grasses and legumes. This type of strip-cropping is not as effective as contour strip-cropping.

Terracing. Terracing with contour farming is more effective as an erosion control practice than strip-cropping, because it positively divides the slope into segments equal to the terrace spacing. With terracing, the slope length to be used in determining the values of LS in the erosion equation is the terrace interval; with strip-cropping or contouring

TABLE 4.12. **Strip Widths Recommended for Contour Strip-Cropping**

Land Slope (%)	Width of Strip (ft)
2–7	88–100
8–12	74–88
13–18	60–74
19–24	50–60

alone, the entire field slope length is used to determine *LS* values.

With terracing and contour strip-cropping, measured soil losses have included only soil moved completely off the field. The soil saved with contour strip-cropping is largely that deposited in the sod strip. With terracing, the deposit is in the terrace channel and may equal up to 90 percent of the soil moved to the channel. This soil is not lost from the field but accumulates in the terrace channel and may result in extra terrace maintenance. If this soil is not moved back upslope in plowing and maintaining the terraces, then a *P* factor for terraces equal to 50 percent of that for contouring is recommended.

If all furrow slices between the terraces were turned upslope periodically, as with a two-way plow, most or all the soil washed into the terrace channel would be effectively moved back up the slope and a factor value based on the off-the-field rate of loss could be safely applied. Limited data indicate the terrace factor in this case should be about 20 percent of that for contouring.

It is logical to assume that the total movement of soil within a terrace interval is equal to that with contouring alone on the same length and percentage of slope. Erosion control between terraces depends upon the crop rotation and other management practices. Therefore if a control level is desired that will maintain soil movement between terraces within the soil-loss tolerance limit, the *P* factor for terracing should equal the contour *P* factor.

However if the erosion equation is used to compute gross erosion for estimates of reservoir sedimentation rates, a terracing *P* factor equal to 20 percent of the contour factor values is recommended.

The erosion control practice factors *P* to use for terracing may be summarized as follows:

1. In determining if the soil loss in the terraced field is within the soil-loss tolerance for that particular soil, use one of the following:
   a. With normal terrace maintenance use the *P* values given in the middle column of Table 4.13.
   b. If a two-way plow is used in terrace maintenance and the furrows are turned upslope between terraces, use the *P* values given in the last column of Table 4.13.
2. To compute the total soil movement between terraces use the *P* value for contouring given in Table 4.10.
3. To compute the soil loss from the terraced field use the *P* value given in the last column of Table 4.13.

Soil-loss tolerance is the amount of soil that can be lost in tons per acre per year and still maintain a high level of productivity over a long period of time. Establishment of tolerances for specific soils and topography has been largely a matter of collective judgment. Both physical and economical factors are considered. For soils in the United States, the

TABLE 4.13. Practice Factor Values for Terracing

| Land Slope (%) | P Values | |
	Normal maintenance	Upslope plowing[a]
1.1–2	0.30	0.12
2.1–7	0.25	0.10
7.1–12	0.30	0.12
12.1–18	0.40	0.16
18.1–24	0.45	0.18

[a] Use this value also in computing the soil loss from the terraced field.

maximum soil-loss rates thus determined range from 1 to 5 tons per acre per year, depending upon soil properties, soil depth, topography, and prior erosion. A deep, medium-textured, moderately permeable soil that has subsoil characteristics favorable for plant growth has a tolerance of 5 tons per acre per year. Such soils with less favorable subsoil would have a tolerance of 4 tons per acre per year. Shallower soils and soils with less desirable physical properties would have a decreasing tolerance. In areas where pollution by sediment is critical, tolerance may be established based on reducing sediment pollution rather than for maintaining soil productivity.

The rainfall erosion index measures only the erosivity of rainfall and associated runoff. Therefore the soil-loss equation does not predict soil loss resulting solely from thaw, snowmelt, or wind. In areas where such losses are significant, they must be estimated separately and combined with those predicted by the equation for comparison with soil-loss tolerances.

APPLICATIONS OF THE SOIL-LOSS EQUATION

The primary purpose of the soil-loss prediction equation is to provide specific and reliable guides to help select adequate soil and water conservation practices for farm fields. Where agricultural lands are a major sediment source, the procedure may also be used to compute agricultural sediment production in predicting rates of reservoir sedimentation. Specific applications of the erosion equation are discussed and illustrated below.

ROTATION AVERAGES

The procedure for computing the expected average annual soil loss from a given cropping system on a particular field is illustrated by the following example:

Example: Assume a field in Fountain County, Indiana, on Russell silt loam, has an 8 percent slope about 200 feet long. The cropping system is a four-year rotation of wheat, meadow, corn, corn with tillage and rows on the contour, and with corn residues disked for wheat seeding and turned under in the spring for second year corn. Fertility and residue management on this farm are such that crop yields are rarely less than 85 bushels of corn, 40 bushels of wheat, or 4 tons of alfalfa-brome hay, and the probability of meadow failure is slight.

Procedure: The first step is to refer to the charts and tables discussed in the preceding sections and to select the values of R, K, LS, C, and P that apply to the specific conditions on this particular field. The value of the rainfall factor R is taken from Figure 4.1. Fountain County, in west central Indiana, lies between iso-erodents 175 and 200. By linear interpolation, $R = 185$. The value of the soil-erodibility factor K is taken from Table 4.3. Russell silt loam is considered equal in erodibility to Fayette silt loam for which the table lists $K = 0.38$. The slope-effect chart (Figure 4.2) shows that for an 8 percent slope 200 feet long, $LS = 1.41$. Figure 4.3 shows that Fountain County is within the geographic area to which erosion index distribution curve 16 applies. For the productivity level and management practices assumed in this example, factor C for a W-M-C-C rotation in area 16 is shown in Table 4.8 to equal 0.119. Table 4.10 shows a practice factor value of 0.6 for contouring on an 8 percent slope, and Table 4.11 indicates that the 200-foot slope is not too long for this factor to be applicable. Therefore under the conditions assumed in this example, $P = 0.6$.

The next step is to substitute the selected numerical values for the symbols in the erosion equation and solve for A. In this example, $A = 185 \times 0.38 \times 1.41 \times 0.119 \times 0.6 = 7.1$ tons of soil loss per acre per year. If planting had been upslope and downslope instead of on the contour, the factor P would have equaled 1.0, and the predicted soil loss for this field would have been $185 \times 0.38 \times 1.41 \times 0.119 \times 1.0 = 11.8$ tons per acre.

CROP-YEAR AVERAGES

The soil losses computed in the example are rotation averages over a long time period. Thus, the heavier losses experienced during the corn years are diluted by trivial losses during the meadow year. Refer again to the solution above, in which the rotation average was 7.1 tons per acre per year. The four-year loss from each complete rotation cycle would average 4×7.1, or about 28.4 tons per acre.

Use of the values in column 8 of Table 4.8 enables computation of the average soil loss for each of the four crop years. Column 8 shows a computed C value of 0.0787 for the wheat period and a C of 0.4764 for the entire four-year period. The average yearly soil loss from wheat in the above example, with contouring, would be $28.4 \times 0.079/0.476$, or 4.7 tons per acre. First year corn including the winter period would average $28.4 \times 0.151/0.476$, or 9.0 tons per acre. The second year corn would average $28.4 \times 0.242/0.476$, or 14.4 tons per acre, and the twenty-month meadow period would average less than 0.5-ton soil loss per acre.

PROBABILITIES OTHER THAN AVERAGE

Because rainfall differs from year to year, the actual value of the factor R also differs from year to year at any given location. Table 4.1 lists 50, 20, and 5 percent probability values of R at key locations. These may be used for further characterization of soil-loss hazards. Fountain County, Indiana (where our example was located), is not listed in the table, but Figure 4.1 shows that the R value there is essentially the same as the R value at Indianapolis. Table 4.1 shows that over a long period the value of the factor R will equal or exceed 225 at Indianapolis in 20 percent of the years. This is $225 \div 185$, or 1.22 times the average value. Returning once more to the example, soil loss from second year corn on the assumed field would be expected to exceed $1.22 \times 14.4 = 17.6$ tons per acre in 20 percent of the years.

INDIVIDUAL STORM SOIL LOSS

The assembled plot data show conclusively that the relation of soil loss to such major factors as slope, cropping, management, and conservation practices is not the same from storm to storm or from year to year, even on the same field under a continuing rotation. In a particular rainstorm, the factor relations are influenced by such variables as antecedent moisture, tillage, tractor and implement compaction, soil crusting by prior rains, and progressive changes in plant cover. Daily soil moisture and temperatures are more favorable to rapid development of good protective cover in some years than in others. The factor values reported in the preceding section and used in the foregoing examples represent average factor relations derived from research measurements over an extended period. Therefore the erosion equation is particularly designed to predict average annual soil loss from any specific field over an extended period.

Predictions of individual storm soil losses will be less accurate, because effects of the minor variations in antecedent conditions cannot be precisely evaluated at this time. However, valuable estimates of single storm losses can be computed.

Procedure: Let R equal the computed erosion index value for the specific rainstorm taken from Table 4.2. Let C in the equation equal the soil-loss ratio shown in Table 4.4, 4.5, or 4.6 for the specific conditions existing on the field at the time of the rain. For example, Table 4.2 shows that a ten-year rain at Indianapolis has an erosion index value of 75 or more. Assume that such a rain occurred about three weeks after planting the second year corn in the preceding example. The existing condition is then described by line 13 of Table 4.4. Since the rain occurred within thirty days after corn planting, the value of C at the time of this particular rain is 48. The value of R is 75. Other values in the equation remain the same as in the first solution. The estimated soil loss from this single rainstorm on the second year corn without contouring is then $RKLSCP = 75 \times 0.38 \times 1.41 \times 0.48 \times 1.0 = 19.3$ tons per acre.

SELECTING CONSERVATION PRACTICES

The soil-loss prediction procedure provides the practicing conservationist with concise, ready reference tables from which he can ascertain for each particular situation encountered which specific land-use and management combinations will provide the desired level of erosion control. A number of possible alternatives are usually indicated. From these the farmer will be able to make a choice in line with his desires and financial resources.

Management decisions generally influence erosion losses by affecting the factor C or P in the erosion equation. The factor L is modified only by terracing. The other three factors—R, K, and S—are essentially fixed so far as a particular field is concerned. When erosion is limited to the maximum allowable, or tolerance, rate, the term A in the equation is replaced by T, and the equation is rewritten in the form: $CP = T/RKLS$. Substituting the locational values of the fixed factors in this equation and solving for CP give the maximum value that the product CP may assume under the specified field conditions. With no conservation practices the most intensive cropping plan that can be safely used on the field is one for which the factor C just equals this value. When a conservation practice such as contouring or strip-cropping is added, the computed value of CP is divided by the practice factor P to obtain the maximum permissible cropping-management factor value. With terracing, the value of $T/RKLS$ is increased by decreasing the value of L.

Since a practicing conservationist usually works within the limits of a single county or other small geographic area, he is usually concerned with only one value of R, one erosion index distribution curve, K and T values for only a few soils, and C values for only a limited number of cropping systems. Therefore the R value for his county, a list of T and K values for the soils in his work area, a few brief tables of pertinent $T/RKLS$ values, and a table of C values for pertinent rotations provide all the information he needs to use this procedure as a guide to selection of conservation practices. He will rarely, if ever, need to solve the equation or to perform computations in the field.

Within a given erosion index distribution area, C values for rotations may be centrally computed for all cropping systems encountered, based on average seeding and harvest dates within that area. The factor for each cropping system needs to be computed for each of several crop productivity levels and for each of several methods of residue management and seedbed preparation. The results are then listed in a table in order of declining magnitude of C, as illustrated in Table 4.9 for area 16.

Example: To illustrate the selection process, we will assume a field in a county having a rainfall factor of 180, located within erosion index distribution area 16. Assume that the soil on this field has a K value of 0.31 and a soil-loss tolerance of 4 tons per acre per year. Past yields on the field have been from 2 to 3 tons of hay and from 60 to 74 bushels of corn per acre. Crop residues will be left over winter and the field will be plowed in the spring.

The land slope averages about 3 percent over the upper half of a total 400-foot slope, but the lower half steepens considerably and ranges from 5 to 7 percent. The field is planted as a single unit. In conservation farming, soil movement from the most vulnerable part of the field should be held below the tolerance limit T. Therefore the gradient of the lower half of the field is the significant percentage of slope for the soil-loss estimate. However, surface runoff from the upper half passes over the lower half. Therefore the overall length is the effective slope length. Thus a slope length of 400 feet and a slope gradient of 6 percent would be used.

Procedure: In the illustration $T = 4$, $R = 180$, $K = 0.31$, $LS = 1.35$ (Fig. 4.2). Substituting these values into the equation $CP = T/RKLS$, we find $CP = 0.053$. If no conservation practice is used, the value of P is 1 and C equals 0.053. Entering Table 4.9 in the column for spring plowing with conventional tillage and 60–74 yield level, we find that the most intensive cropping system that can be used and keep C below 0.053 is C-O-M-M-M.

With contour farming, P is equal to 0.50 (Table 4.10), so C equals 0.11. From Table 4.9 we find that the most intensive rotation would be C-C-O-M-M-M. However, as indicated in Table 4.11, the effective slope length for contouring is only 300 feet. The soil loss on the lower 100 feet of the slope would exceed the tolerance limit so the farmer may wish to consider using strip-cropping or terracing.

A system of terraces would reduce the slope length to approximately 100 feet. The new value of LS from Figure 4.2 is 0.65 which, substituted in the equation above, gives a value of CP equal to 0.11. If normal terrace plowing and maintenance is practiced, P would have a value of 0.25. (See Table 4.13.) Solving for C gives a value of 0.44. From Table 4.9, we find that continuous corn, $C = 0.43$, could be used.

Example: Assume that the average yearly incomes per acre above costs for crops in this area are corn—$60, wheat—$35, oats—$15, and meadow—$20, and that the average yearly investment and maintenance cost for terracing is $4 per acre. The results of the above illustrative example can be summarized as follows:

Erosion control practice	Most intensive rotation	Average yearly income	Practice expense	Net income
None	C-O-M-M-M	$27	$0	$27
Contouring	C-C-O-M-M-M	33	0	33
Terracing	Continuous corn	60	4	56

From an analysis of the above data, the farmer would be able to select the combination of cropping system and conservation practice that would best fit his needs and result in the highest income.

Procedure: To determine the soil losses from continuous corn in the example, substitute the C value of 0.43 in the erosion equation, $A = RKLSCP$, and solve for (1) the total soil loss between terraces, $A = 180 \times 0.31 \times 0.65 \times 0.43 \times 0.50 = 7.8$ tons per acre per year, and (2) the soil loss from the field, $A = 180 \times 0.31 \times 0.65 \times .043 \times 0.10 = 1.6$ tons per acre per year. The difference $(7.8 - 1.6)$ of 6.2 tons per acre per year represents the amount of soil that is deposited in the terrace channels. This could be reduced by improving the general fertility level of the field, by using minimum tillage, or by using other management practices which would reduce erosion between terraces. If upslope plowing were used, this soil would be moved out of the channel and back up the slope.

IMPROVEMENTS IN THE SOIL-LOSS EQUATION

Research is being continued to evaluate additional soils, cropping-management practices, and topographic factors to improve the usefulness of the soil-loss equation. These research data are being collected and analyzed by the USDA-ARS Runoff and Soil-Loss Data Center at Purdue University.

PROBLEMS

4.1. Compute the cropping-management factor C for a rotation of C-O-M in southwestern Iowa. Assume that conventional tillage is used, that an average production of 4 tons of hay and 90 bushels of corn per acre is expected, that the meadow is a mixture of grass and legume, and that residues are left on the field. Plow for corn May 1, plant corn May 20, harvest corn October 15, disk stalks and seed oats March 15, and harvest oats July 15.

4.2. Assume that the cropping system in problem 4.1 is used in southwestern Iowa on a Marshall silt loam soil. The field has a slope of 6 percent and a slope length of 500 feet. All farming operations are on the contour.
   a. Compute the average soil loss from the rotation.

b.   Compute the soil loss expected from the time the land is plowed for corn until the cornstalks are disked for oats.

c.   Compute the soil loss to be exceeded one year in twenty years during the period specified in (b). Use the *EI* value for Des Moines, Iowa.

d.   Compute the tons of soil loss to be exceeded once in twenty years from an individual storm if it were to occur on June 30 during the corn crop.

4.3.   A field in west central Illinois has a 6 percent slope and a slope length of 500 feet. The soil erodibility factor is 0.33. A cropping system of C-C-O-M is to be used. (Use *C* values from Table 4.9.) Average yields of 4 tons of hay and 90 bushels of corn per acre can be expected.

a.   What will be the average annual soil loss from the field if conventional tillage is used and residues are left over winter—

(1) with all operations up and down the slope? (2) with all operations on the contour? (3) with all operations on the contour with strip-cropping? and (4) with all operations on the contour with terraces (120-foot spacing)?

b.   What will be the average annual soil loss from the area between terraces with conventional tillage?

c.   What will be the average annual soil loss from the field if conventional tillage is used with all operations on the contour, and the fertility and management levels decrease so that average yields of 2.5 tons of hay and 65 bushels of corn per acre can be expected?

4.4.   The farmer who owns the field referred to in problem 4.3 plans to use such crop management and erosion control practices that his soil loss will be equal to or less than 4 tons per acre per year. What cropping system will provide him the highest net income, with conventional tillage?

a.   If he farmed up and down the slope?

b.   If he farmed on the contour?

c.   If he farmed on the contour with strip-cropping?

d.   If he farmed on the contour with terraces and used a conventional moldboard plow?

e.   If he farmed on the contour with terraces, used a two-way plow, and turned all furrows upslope?

REFERENCE

1.   Wischmeier, W. H., and Smith, D. D. Predicting rainfall-erosion losses from cropland east of the Rocky Moutains. USDA Agr. Handbook 282, May 1965. (May be obtained from the Superintendent of Documents, USGPO, Washington, D.C. 20402, 30 cents.)

5 ~ Rainfall and Runoff

*A study of rainfall and runoff is included
in the broad field of hydrology which
encompasses the study of water as it
occurs in the atmosphere as well as on and
under the surface of the earth.*

THE earth's water circulatory system is known as the hydrologic cycle
(Fig. 5.1) and this study of it will begin with precipitation which may occur as rain, snow, hail, or sleet. Some of this precipitation evaporates before reaching the ground. Some is intercepted by vegetation, buildings,
and other objects and is evaporated later. A part infiltrates into the soil
to be disposed of in the following ways: some is retained in the upper
layers of the soil where it may be taken into the root systems of plants
and transpired or moved to the surface and evaporated; some moves laterally in the soil to surface drainageways; some moves downward and becomes part of the groundwater supply where it may be stored or will
eventually move into streams or the ocean. The remainder—that part
which is not evaporated, does not enter the soil, or is not ponded on
the soil surface—moves over the surface as runoff; and most of it eventually reaches the ocean. Moisture evaporated from the oceans may be
precipitated and returned to the ocean, or it may be carried by
winds and precipitated over the land areas. In erosion control we are
primarily concerned with precipitation in the form of rain and snow.

In northern climates a considerable proportion of the precipitation is in the form of snow, which in most cases is beneficial to the land.
Snow that falls on unfrozen land usually thaws slowly enough to supply
water to the soil without causing erosion. A snow that falls on frozen
soil and stays there for a long period of time may serve as an insulating
layer, permitting the heat from below to reduce its thickness or perhaps
to thaw it completely. Moisture from the snowmelt can then penetrate
the soil. If the snow melts while the ground is still frozen, a larger
percentage of the water will be lost as runoff; and if the soil surface is

69

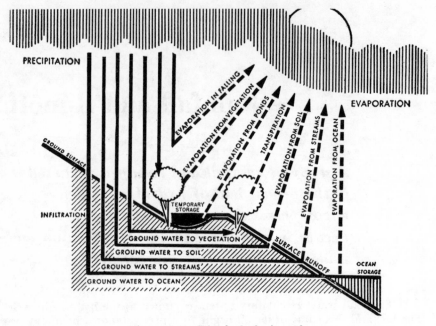

Figure 5.1. The hydrologic cycle

thawed, serious erosion may result. If a rain occurs at this time, erosion may be even more severe. When accompanied by wind, the snow is often piled up in huge drifts, the size depending upon the extent to which obstructions reduce the velocity of the wind and cause it to drop its load of snow. If these drifts thaw rapidly, they may result in serious gullying.

RAINFALL

Precipitation in the form of rain produces the greatest amount of runoff and erosion. Some rains are highly beneficial, supplying the needs of microorganisms in the soil and of crops on the land. Others cause damage far in excess of any benefits. These damaging rains may come at a rate that is much greater than the infiltration capacity of the soil and may cause runoff and erosion. They may come at a slower rate, but continue after the soil has become saturated. Almost all the rain that falls after this point is reached must leave the land as surface runoff. The soil is held together very loosely in its saturated condition and is easily detached. Intense rains may come at a time when the land is being prepared for seeding. Infiltration is rapid at first but as soon as the soil pores are reduced by the puddling and pounding action of the raindrops and clogged with the fine material from the muddy water, infiltration is reduced and runoff increased during the remainder of the rain.

The principal characteristics of rains that affect runoff and erosion are intensity, duration, distribution of rainfall intensity throughout the storm, frequency of occurrence, seasonal distribution, and areal distribution.

INTENSITY AND DURATION

We know from experience that rains of high intensity generally last for fairly short periods of time and cover relatively small areas. Storms covering large areas are usually less intense but last for a longer time. Intense rains of varying duration occur from time to time over much of the country. However, the probability of occurrence varies with the locality and increases as the length of time increases. For example, a more intense rain could be expected in the next 100 years than in the next ten years.

Information on rainfall to be expected in the United States may be found in Hershfield (2). Available maps show rainfall durations of one-half, one, two, three, six, twelve, and twenty-four hours and return periods of one, two, five, ten, twenty-four, fifty, and one hundred years (2).

Figure 5.2 can be used to determine the maximum rainfall to be expected in a ten-year period for a thirty-minute duration storm. Figure 5.3 gives the maximum rainfall to be expected in central Missouri for

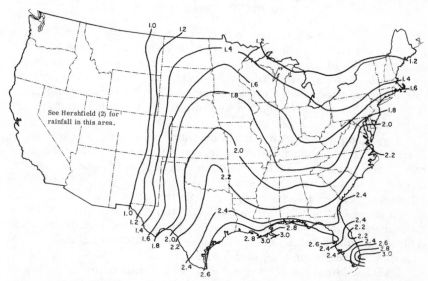

Figure 5.2. Maximum amount of rainfall, in inches, to be expected in a 10-year period for a 30-minute duration storm. Data from D. M. Hershfield, Rainfall frequency atlas of the United States, U.S. Weather Bureau Tech. Paper 40, May 1961.

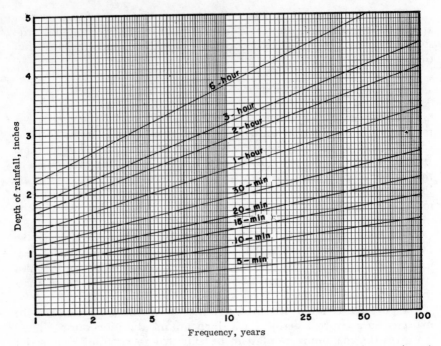

Figure 5.3. Maximum rainfall to be expected in central Missouri for rains of various frequencies of occurrence.

rains of various durations and frequencies of occurrence. Since the relationship between maximum rainfall and duration and frequency of occurrence are rather consistent over a wide area, these two figures can be used to estimate, with a reasonable degree of accuracy, the maximum rainfall to be expected for various durations and frequencies in areas of the United States east of the Rocky Mountains. For more accurate estimates, see Hershfield (2).

For example, determine the maximum rainfall to be expected in north central Oklahoma in a fifty-year period from a rain lasting one hour. (1) From Figure 5.3 the maximum depth of rainfall to be expected in central Missouri in a fifty-year period from a rain lasting one hour is 3.1 inches. This is 1.63 times greater than the 1.9 inches to be expected from a ten-year, thirty-minute rainfall. (2) From Figure 5.2 the maximum rainfall to be expected from a ten-year, thirty-minute rainfall in north central Oklahoma is 2.2 inches. The fifty-year, one-hour rainfall will be 1.63 times greater than this, or 3.6 inches.

Rainfall intensities are usually expressed in inches per hour. Knowing the amount of rain that falls in a given time, the intensity can be computed. For example, if 1.9 inches of rain fall in thirty minutes, the intensity would be 3.8 inches per hour; if 0.9 inch of rain falls in five minutes, the intensity would be 10.8 inches per hour.

A rainfall seldom occurs at a uniform rate throughout the duration of the storm. If the rainfall intensity is greatest at the beginning of a rain, it is classed as an advanced storm. If the intensity is greatest near the middle of the rain, it is classed as an intermediate storm. If the intensity is greatest at the end of a rain, it is classed as a delayed storm. The highest rate of runoff is usually produced by a delayed storm since the highest intensity occurs when the rate of infiltration is least.

FREQUENCY AND DISTRIBUTION

In addition to the intensity and duration, the frequency of occurrence and seasonal distribution of intense rains are extremely important factors affecting runoff and erosion. If intense rains occur during the time that a seedbed is being prepared, during planting time, or when crop canopies are being established, they usually cause considerable runoff and erosion. These same rains occurring at other periods of the crop cycle may cause little or no damage. In planning crop rotations, it is important to know when to expect these intense rains so that, if possible, effective crop cover can be established during these periods.

The rainfall factor R (Ch. 4) gives an indication of the frequency of intense rains. Those areas having the higher average annual value of the rainfall factor R (Fig. 4.1), would in most cases have a greater number of intense rains. The seasonal distribution of the intense rains can be obtained from Table 4.7.

For example, in southwest Iowa the average annual value of R is approximately 175. The percentage of this value that occurs in each month is as follows: January—1, February—0, March—2, April—4, May—12, June—28, July—18, August—16, September—12, October—5, November—1, and December—1. It is noted that May and June account for 40 percent of the total yearly value. Corn and soybeans are grown extensively in this area. During May and June these crops offer little protective crop canopy. Twenty-eight percent of the average annual value of R occurs during August and September. This is the period when land seeded for small grain or pasture has little protective cover. Because of excessive rainfall during these periods, runoff and erosion on those fields where erosion control practices are not used are excessive.

The average annual value of R for southern Illinois is 200, which is greater than the value for southwestern Iowa. The percentage of this value occurring each month is as follows: January—3, February—6, March—8, April—9, May—11, June—13, July—14, August—13, September—7, October—6, November—5, December—5. This distribution is much more uniform throughout the year with only 24 percent occurring in May and June and 20 percent in August and September. Even though the yearly value of R is higher, the erosion is less in this area because more of the rains occur when the soil is protected by a vegetative cover.

The hazards involved in the production of crops in other areas may be evaluated by determining the vulnerable periods in the production of the crop and the probability of intense rains occurring during these periods.

If possible, cropping systems should be selected so that protective cover is on the ground when intense rains can be expected. If this is not possible, then additional erosion control measures such as contour farming, strip-cropping, or terracing will be needed to protect the land.

The rainfall data in Figures 5.2 and 5.3 refer to rainfall at a given point. The rainfall distribution over a given area will vary from this value, for seldom is rainfall distributed evenly over a large area. The areal distribution of rainfall over a watershed affects the rate of runoff to be expected.

RUNOFF

In the study of the hydrologic cycle, we found surface runoff to be that part of precipitation which flows over the ground surface and through channels to larger streams. Another part of precipitation infiltrates into the soil and moves laterally to surface drainageways. This is subsurface flow or interflow. In the design of structures for the control of erosion and conservation of water on small areas we are primarily concerned with surface runoff. We must be able to predict the maximum rate of runoff to be expected through structures that are to convey runoff and the total amount of runoff that can be expected to flow into structures built for storage.

Before surface runoff can occur, precipitation must be in excess of that required for evaporation, interception, infiltration, surface detention, and storage. The amount of water intercepted and evaporated during an intense rain of long duration is so small that it will have little effect in reducing surface runoff. On the other hand a light rainfall may be almost entirely intercepted by a dense vegetative growth.

INFILTRATION

Infiltration is the movement of water into the soil surface. The higher the rate of infiltration the lower the rate of surface runoff and erosion. Movement of water through the soil is brought about by the forces of gravity and capillarity. Movement through large pores when the soil is saturated is primarily by gravity whereas movement through unsaturated soil is primarily by capillarity.

During a rainstorm the maximum infiltration rate occurs at the beginning of the rain and usually decreases very rapidly as structural changes in the surface soil occur. If the rain continues, the infiltration rate gradually approaches a minimum value. This minimum value is determined by the rate at which water can enter the surface layer and by the rate at which it can be transmitted through the soil profile.

The size and arrangement of pore spaces determines to a large extent the infiltration rate of a soil. Sandy soils having large pore spaces may be expected to have a higher infiltration rate than the silts and clays which have relatively small pore spaces. The infiltration rate is also affected by varying texture in the soil profile. If a sandy soil is underlain at some depth by a relatively impermeable clay layer, we would expect the infiltration rate to be high until the sandy layer became saturated, at which time it would be reduced to that of the clay layer. If the clay soil were at the surface, the infiltration rate at the beginning of the rain would be less, and the variation in the infiltration rate during the rain would not be as great.

The moisture content of the soil at the beginning of the rain also affects the infiltration rate. Colloidal material in the soil tends to swell when wetted, thereby reducing both the size of pore space and rate of water movement. Soils with a high content of colloidal material tend to crack when dry, resulting in a high infiltration rate until the cracks are filled. The effect of soil moisture on infiltration is greatest on those soils with a large percentage of colloidal material. Soil moisture is usually higher in the spring than during the summer. Hence those practices which tend to increase the possibility for increased infiltration are more effective in reducing runoff during the summer months.

The extent of soil aggregation is another factor affecting infiltration. If finer soil particles are well aggregated, the pore spaces between the aggregates are larger, providing for a higher infiltration rate. Soil management practices that improve the physical condition and granulation of the soil reduce the runoff and erosion resulting from most rains.

Cultivation has the effect of temporarily loosening surface soil and increasing infiltration. However if the surface is not protected by vegetation or mulches, rain and wind soon consolidate the surface and reduce the infiltration rate. Special methods of cultivation may be used under certain conditions to increase infiltration and prevent surface sealing. So-called "subsurface cultivation" stirs the ground beneath a surface mulch. This results in a higher infiltration rate but creates other problems in crop production. Such practices as listing and basin listing provide for increased surface storage and some increase in soil-water contact. Depth of cultivation is also an important factor in increasing the infiltration rate. Tillage practices that pack the soil, such as rolling or dragging, tend to decrease the infiltration rate. Cultivation on the contour delays runoff and gives more time for the infiltration process.

The most important factor affecting the infiltration rate is the vegetative cover that is on the soil at the time of the rain. If an intense rain comes when the soil is not protected by a vegetative cover or by a mulch, it packs and seals the surface layer, reducing infiltration. However if such a rain comes when there is a vegetative cover, the soil remains permeable and has a much higher infiltration rate. The great value of soil improvement practices in erosion control comes from the ability of the improved soil to produce a more effective vegetative cover.

If the rate of rainfall does not exceed the rate of infiltration there is no runoff and no erosion problem. When the rate of rainfall exceeds the infiltration rate, surface depressions begin to fill and then overflow to produce surface runoff. After the surface depressions have filled and surface runoff starts, *the rate of runoff is determined by the difference between the infiltration rate of the soil and the rainfall intensity.*

The maximum, or peak, rate of runoff from small areas usually results from *a rain that covers the area uniformly and lasts long enough that there is a concentration of runoff from all parts of the watershed.* The time required to have an accumulation of runoff from all parts of the watershed is called the *time of concentration.* Thus if the time of concentration of a watershed is short, a rain must last only a short time to produce maximum runoff. If it is longer, the short, intense rain does not result in a concentration of runoff from all parts of the watershed nor is it likely that the intense rain will cover the entire area; hence a maximum rate of runoff will not result.

Two typical watersheds in central Missouri are shown in Figure 5.4. The time of concentration for the 30-acre watershed is ten minutes, so the maximum rate of runoff is produced by rainfall lasting ten minutes. The depth of rainfall expected in ten minutes from a ten-year frequency storm is 1.08 inches (Fig. 5.3). The intensity is 6.5 inches per hour. The time of concentration for the 100-acre watershed is

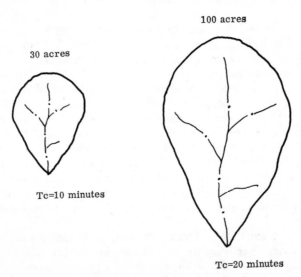

Figure 5.4. Typical watersheds to be used in studying runoff. (Tc = time of concentration.)

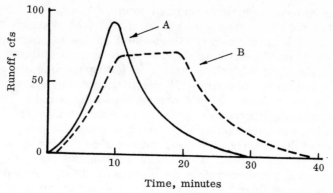

Figure 5.5. Hydrographs of runoff from the 30-acre watershed in Figure 5.4.

twenty minutes, so the maximum rate of runoff is produced by a rainfall lasting twenty minutes. The depth of rainfall expected in twenty minutes from a ten-year frequency storm is 1.6 inches (Fig. 5.3). The intensity is 4.8 inches per hour.

THE HYDROGRAPH

The fluctuation in runoff from an area plotted graphically with respect to time is called a hydrograph. The rate of runoff is usually expressed in cubic feet per second (cfs). The hydrographs of runoff from the 30-acre watershed are shown in Figure 5.5 and from the 100-acre watershed in Figure 5.6. The curves marked *A* are hydrographs

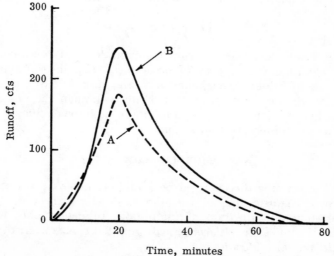

Figure 5.6. Hydrographs of runoff from the 100-acre watershed in Figure 5.4.

resulting from a rainfall lasting ten minutes having an intensity of 6.5 inches per hour. The curves marked *B* are hydrographs resulting from a rainfall lasting twenty minutes and having an intensity of 4.8 inches per hour. In each case the rainfall intensity is uniform throughout the rain. The 6.5-inch-per-hour rain lasting ten minutes produced a peak runoff of 94 cfs from the 30-acre watershed, but did not produce maximum runoff from the 100-acre watershed since it did not last long enough to result in a concentration of runoff from the entire watershed. The 4.8-inch-per-hour rain lasting twenty minutes did produce a peak runoff from the 100-acre watershed, but not from the 30-acre watershed. At the end of ten minutes, runoff was accumulating from all parts of the 30-acre watershed, and since there was no additional area contributing to the runoff there was a sharp reduction in the rate of rise in the hydrograph after that time. However, the runoff rate continued to increase slightly as long as the rain continued because of the reduction in the infiltration rate and the resulting increase in runoff. The peak rate of runoff from the 30-acre watershed was 94 cfs, or 3.1 cfs per acre. The peak rate of runoff from the 100-acre watershed was 245 cfs, or 2.45 cfs per acre. The higher peak rate per acre on the smaller watershed is a result of the more intense rain.

In the design of larger water control structures the hydrograph of the runoff flowing to them is essential. However, many smaller structures and conservation practices are designed to carry the peak rate of runoff expected from a watershed. In their design the peak rate of runoff is required but usually the complete hydrograph is not.

The peak rate of runoff and the shape of the hydrograph to be expected from a watershed are influenced by the combination of a number of climatic and watershed characteristics.

CLIMATIC FACTORS

Climatic factors affecting runoff are the form of precipitation, its intensity, duration, frequency of occurrence, areal distribution, distribution of intensities throughout the storm, direction of storm movement, and the degree of saturation of the soil when the precipitation occurs. Evaporation and transpiration are additional climatic factors which have a minor effect on runoff.

WATERSHED CHARACTERISTICS

In determining the expected peak rate of runoff we are concerned with those characteristics of the watershed which affect the time of concentration and the rate of infiltration. In determining the total amount of runoff we are concerned with those characteristics which affect the rate of infiltration.

Size. The rate of runoff increases as the size of the watershed increases. However, the rate per unit of area decreases as the size of the watershed increases because the larger watershed has a longer time of concentration, and a longer rain is required to produce maximum runoff. The longer rain will be less intense and there will be less difference between the rate of rainfall and rate of infiltration; hence the rate of runoff per unit of area will be less.

Drainage basins may be classified as large or small not only on the basis of area, but also on the basis of their hydrologic characteristics. A small basin is one in which the runoff rate is influenced to a great extent by high-intensity rainfalls of short duration and by land management practices. A large basin, on the other hand, is one in which the runoff rate is influenced more by channel shape, size, and channel storage than it is by land management practices (3).

Soil Type. The permeability of the soil affects the infiltration rate more than it affects the time of concentration.

Slope. On steep slopes water runs off faster, decreasing the time of concentration. The rate of infiltration is possibly somewhat less on the steeper slope. Higher rates of runoff can thus be expected as the slope of the land increases.

Shape. The peak rate of runoff is affected by the maximum distance runoff must travel in reaching the discharge point. Long narrow watersheds have a longer time of concentration than more compact watersheds of the same size and hence a lower rate of runoff.

The location of terraces or diversions and their direction of flow relative to the point of measurement affects the time of concentration and the resulting peak rate of runoff from a watershed. This is illustrated by actual hydrographs resulting from short intense rains on the Burge Branch Watershed near Arrow Rock, Missouri.

Figure 5.7 shows the watershed in 1959 when very few terraces had been built. The resulting hydrograph is typical; it has only a single peak and the flow from the watershed was reduced to a trickle at the end of three hours. Figure 5.8 shows the watershed in 1961 after a number of terraces had been built. It will be noted that the terraces in the upper portion of the watershed drain away from the gaging station and those in the lower portion drain toward the station. The resulting hydrograph is typical. It has two peaks, the first resulting from runoff from those terraces near the gaging station and the second from runoff from those in the upper portion of the watershed draining away from the discharge point. The flow from the watershed also continued over a longer period of time.

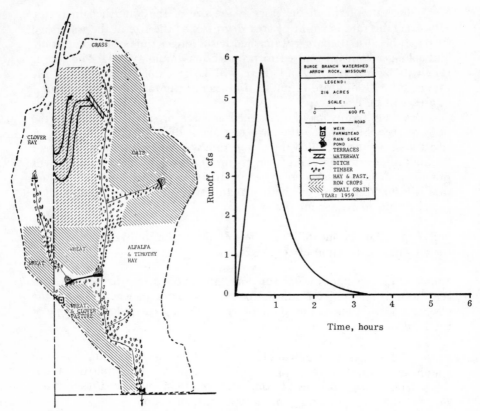

Figure 5.7. The Burge Branch Watershed in 1959 and a hydrograph resulting from a rain of 0.46 inch falling in sixteen minutes.

Vegetative Cover. The density of the ground cover affects the infiltration rate more than it affects the time of concentration.

Contour Farming. Practices such as contour farming and strip-cropping which retard the flow of water are effective in increasing both the amount of infiltration and the time of concentration. However since the length of time that the flow can be retarded by these practices is limited, their effect on increasing the time of concentration and reducing peak rate of flow is greatest on small watersheds and decreases rapidly as the size increases. Their effect on increasing infiltration is also greatest on small watersheds where the maximum rates of runoff usually occur during the growing season when the soil is dry enough that a delay in runoff results in increased infiltration.

Surface Storage. Any water that is detained or impounded on the watershed reduces the rate and amount of runoff from the watershed. Storage

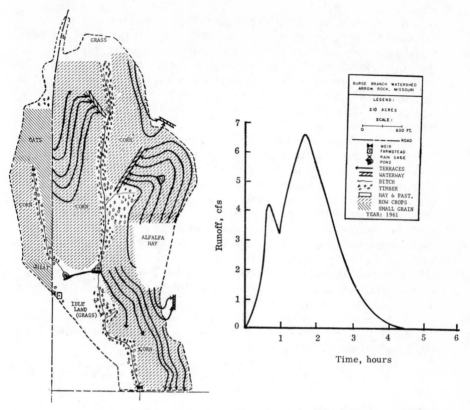

Figure 5.8. The Burge Branch Watershed in 1961 and a hydrograph resulting from 0.70 inch of rain falling in seventeen minutes.

may be in natural depressions or lakes or it may be created by the construction of ponds, stabilization and flood retardation structures, reservoirs, and terraces. Detention storage is provided in terrace channels and in crop rows which are planted on the contour.

The amount that storage reduces peak rates of runoff depends on the percentage of the watershed area that drains into the structure and the amount of storage provided. For example, level terraces constructed on an entire watershed impound the surface runoff and practically eliminate it from the watershed for all but the most extreme rains which would exceed the storage capacity of the terraces.

PEAK RATES

In the design of water management structures, it is essential that an estimate be made of the peak rate of runoff to be expected from the area draining into the structure for a given design period. The design period is selected by comparing the loss that would occur if the runoff exceeded the capacity of the structure to the cost of providing the additional capacity. Terraces and terrace outlets are usually constructed for the peak rate of runoff to be expected in a ten-year design period. Diversion channels and stabilization structures vary widely in size and in the damage which would result if they overflowed; consequently, the design period selected may vary from a period of 10 years to 50 years or longer.

Many methods have been developed for estimating the peak rate of runoff to be expected in a given design period (3). Empirical formulas have been widely used in the past but in most cases they evaluated only a limited number of the factors affecting runoff. Tables and curves giving measured runoff from similar watersheds in the immediate area prove valuable if the measurements have been continued over a relatively long period and the gaged watershed is quite similar in all respects to the watershed under consideration. Prediction of runoff rates from an analysis of actual runoff measurements from the watershed being considered is probably the most accurate method but seldom, if ever, are runoff records of adequate duration available for the small watersheds usually considered in erosion control work.

A prediction equation, similar to the soil-loss equation, has been developed by the author for estimating the peak rate of runoff from small watersheds. The equation is

$$Q = Q_T \times L \times I \times T \times S \times V \times C \times P \times F$$

where Q = peak rate of runoff, cfs
$\quad Q_T$ = peak rate of runoff from a watershed with a specific set of watershed conditions (Table 5.1)
$\quad L$ = watershed location factor
$\quad I$ = soil infiltration factor
$\quad T$ = topographic factor
$\quad S$ = watershed shape factor
$\quad V$ = vegetative cover factor
$\quad C$ = contour farming factor
$\quad P$ = surface storage factor
$\quad F$ = runoff frequency factor

Table 5.1 gives peak rates of runoff to be expected under a given set of conditions in an average ten-year period from various sizes of watersheds located along line 1.0 in Figure 5.9. If the watershed for which the peak rate of runoff is being estimated has conditions differing from those in Table 5.1, the values from the table are multiplied by an appropriate factor for the conditions that exist for the watershed being considered.

TABLE 5.1. Peak Rates of Runoff to Be Expected from Watersheds

Acres	Runoff, cfs	Acres	Runoff, cfs	Acres	Runoff, cfs
5	19	90	225	300	570
10	36	95	235	320	598
15	52	100	245	340	626
20	67	110	265	360	654
25	81	120	285	380	682
30	94	130	304	400	710
35	107	140	322	420	736
40	120	150	340	440	762
45	132	160	356	460	788
50	144	170	372	480	814
55	155	180	388	500	840
60	166	190	404	520	864
65	175	200	420	540	888
70	185	220	450	560	912
75	195	240	480	580	936
80	205	260	510	600	960
85	215	280	540		

Note: The following conditions apply: ten-year frequency runoff; located along line 1.0, Figure 5.9; soils with an average infiltration rate; average land slope—8 percent; a typical shape; in row crop, with crop rows planted across the slope without terraces; and no appreciable surface storage.

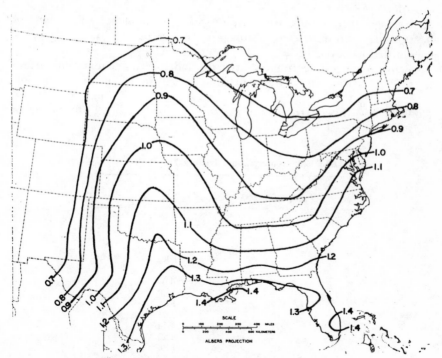

Figure 5.9. Values of the location factor, L.

Location Factor, **L.** The rate of runoff is affected by the intensity of rainfall expected on the watershed. For locations other than along line 1.0 (Fig. 5.9), the value in Table 5.1 is multiplied by the appropriate factor from Figure 5.9. Because of the wide variation in rainfall intensity expected in the mountainous areas of the western United States, the values of L for these areas are not shown on the map. However, the value of L for these areas may be computed by determining the maximum depth of rainfall in inches expected during a ten-year period for a rainstorm of thirty-minute duration and dividing this value by 1.9.

Soil Infiltration Factor, **I.** The rates of runoff given in Table 5.1 are those expected from a soil with an average infiltration rate. If soil conditions differ from this, the value from Table 5.1 should be multiplied by an appropriate factor I determined from the soil infiltration rate. For a listing of soils according to their infiltration rate, see references 7 and 8. With an infiltration rate that is—

1. Very high (coarse-textured soils throughout), $I = 0.8$.
2. Above average (medium- to moderately coarse-textured surface soils and well-drained subsoils as evidenced by bright or nonmottled colors), $I = 0.9$.
3. Average (medium-textured surface soils and moderately fine-textured subsoils with restricted drainage as evidenced by having gray or mottled colors), $I = 1.0$.
4. Below average (medium-textured to moderately fine-textured surface soils with claypan or fragipan subsoils within 12 inches of the surface), $I = 1.1$.
5. Very low (fine-textured surface soils or soils eroded into claypan subsoils), $I = 1.2$.

TABLE 5.2. Topographic Factors *T*

Average Land Slope (%)	*T*
1	0.65
2	0.72
3	0.78
4	0.83
5	0.88
6	0.92
7	0.96
8	1.00
9	1.04
10	1.07
12	1.14
14	1.20
16	1.26
18	1.32
20	1.37

Topographic Factor, **T.** The rates of runoff given in Table 5.1 are those expected from land averaging 8 percent slope. If the average land slope on a watershed differs from this, the value from Table 5.1 is multiplied by an appropriate factor selected from Table 5.2.

Watershed Shape Factor, **S.** The shape of a watershed may be described by the maximum distance the runoff must travel in reaching the discharge point, including the distance that runoff travels in terraces and diversions. Locate this distance in Table 5.3 opposite the watershed of the appropriate size and determine the shape factor S at the bottom of Table 5.3. Multiply the value from Table 5.1 by this shape factor.

Vegetative Cover Factor, **V.** The rates of runoff given in Table 5.1 are expected if the entire watershed is in row crops. If different covers are involved, the value in Table 5.1 should be multiplied by an appropriate factor selected from the following: urban areas, $V = 2.0$; farmsteads and suburbs, $V = 1.2$; small grain of good quality, $V = 0.8$; small grain of poor quality, $V = 0.9$; pasture of good quality, $V = 0.6$; pasture of poor quality, $V = 0.8$; meadow of good quality, $V = 0.5$; meadow of poor

TABLE 5.3. Watershed Shape Factors S

Acres	Maximum Distance Traveled by Runoff (ft)										
5				400	500	600	700	800	900	1,100	1,300
10		500	550	600	700	800	950	1,100	1,300	1,600	1,900
20	600	700	800	900	1,000	1,200	1,400	1,600	1,900	2,300	2,800
30	800	900	1,000	1,200	1,400	1,600	1,900	2,200	2,600	3,100	3,600
40	1,000	1,100	1,200	1,400	1,600	1,900	2,200	2,600	3,100	3,600	4,200
50	1,100	1,200	1,400	1,600	1,800	2,100	2,400	2,900	3,400	3,900	4,600
60	1,300	1,400	1,600	1,800	2,100	2,400	2,700	3,200	3,700	4,200	5,000
70	1,400	1,600	1,800	2,000	2,300	2,700	3,100	3,600	4,100	4,600	5,400
80	1,500	1,700	1,900	2,100	2,500	2,900	3,300	3,800	4,400	5,000	5,800
90	1,700	1,800	2,000	2,300	2,700	3,100	3,600	4,100	4,600	5,400	6,200
100	1,800	1,900	2,200	2,500	2,900	3,300	3,800	4,400	5,000	5,700	6,600
120	1,900	2,100	2,400	2,800	3,200	3,600	4,100	4,700	5,400	6,200	7,100
140	2,100	2,400	2,700	3,100	3,600	4,000	4,500	5,100	5,900	6,700	7,600
160	2,300	2,600	3,000	3,400	3,800	4,300	4,800	5,500	6,300	7,100	8,000
180	2,500	2,900	3,300	3,700	4,100	4,600	5,200	5,900	6,700	7,500	8,400
200	2,700	3,100	3,500	3,900	4,300	4,800	5,500	6,200	7,000	7,900	8,800
250	3,200	3,600	4,000	4,400	4,900	5,500	6,300	7,100	7,900	8,700	9,600
300	3,700	4,100	4,500	5,000	5,500	6,200	7,000	7,700	8,500	9,400	10,400
350	4,100	4,500	4,900	5,400	6,100	6,800	7,500	8,200	9,000	10,000	11,200
400	4,500	4,900	5,400	6,000	6,700	7,400	8,200	9,000	9,900	10,800	12,000
450	4,900	5,400	5,900	6,500	7,200	7,900	8,700	9,600	10,600	11,600	12,800
500	5,400	5,900	6,500	7,100	7,700	8,400	9,200	10,100	11,100	12,200	13,600
550	5,900	6,400	7,000	7,600	8,300	9,000	9,800	10,700	11,600	12,900	14,300
600	6,300	6,900	7,500	8,100	8,800	9,600	10,400	11,300	12,400	13,700	15,100
Shape factor S											
	1.25	1.20	1.15	1.10	1.05	1.00	0.95	0.90	0.85	0.80	0.75

quality, $V = 0.7$; timber of good quality, $V = 0.4$; timber of poor quality, $V = 0.6$.

Contour Farming Factor, **C.** The rates of runoff given in Table 5.1 are those expected from a watershed on which the majority of the crop rows are planted across the slope but not necessarily on the contour. If the entire watershed is farmed parallel to contour guidelines or terraces, the grade in the crop rows does not exceed 2 percent, and the crop rows discharge into grassed waterways, the value in Table 5.1 is multiplied by an appropriate factor (C): with 0–10 acres in the watershed, $C = 0.95$; 11–40 acres, $C = 0.96$; 41–100 acres, $C = 0.97$; 101–200 acres, $C = 0.98$; and 201–600 acres, $C = 0.99$.

If only a portion of the watershed is farmed on the contour, the factor should be increased. For example, if 30 acres on a 40-acre watershed are farmed on the contour, the contour factor C would be 0.97.

Surface Storage Factor, **P.** Any water that is detained or impounded in the watershed reduces the peak rate of runoff. Values in Table 5.1 are peak rates of runoff expected from a watershed with no appreciable surface storage. If water is impounded on the watershed, the value from Table 5.1 is multiplied by an appropriate factor. The storage factor may vary, depending on the percentage of the watershed that drains into the impoundment, the amount of storage provided, and the location of the impoundment.

Because of the wide variations in types of impoundments and the many possible locations on the watershed, it is impossible to give specific values for the storage factor that would be applicable for all situations. The following general principles may be helpful in evaluating the storage factor:

1. Impoundments with only an emergency side spillway do not reduce the peak rate of runoff an appreciable amount, particularly if they are expected to be full when the design storm occurs.
2. Impoundments with a principal spillway at a lower elevation than the emergency spillway reduce the peak rate of runoff. The amount of reduction depends on the percentage of the watershed that drains into the impoundment, the rate of flow through the principal spillway, the amount of storage provided between the principal and emergency spillway, and the capacity of the emergency spillway.
3. Impoundments located in the upper part of a watershed give the greatest reduction in the peak rate of runoff, those in the lower part give the least reduction, and impoundments throughout the watershed have an intermediate effect.
4. The effect of level terraces on the peak rate of runoff depends on the percentage of the watershed that is terraced and the amount of storage provided.

5. The effectiveness of natural storage such as lakes, swamps, and sink-holes depends on their location on the watershed, the percentage of the watershed draining into them, and the amount of storage available when the design storm occurs.

In order to evaluate the storage factor on any given watershed correctly, the design storm must be flood-routed through the impoundments on the watershed. Since all the information needed to flood-route the storm accurately is seldom available, the following procedure to determine the effects of impoundments on the peak rate of runoff is suggested:

1. Determine the peak rate of runoff expected for the design period from the area below the impoundments.
2. Determine the expected peak rate of runoff from the spillways of the impounding structures for the design storm. The procedure to follow is given in Chapter 9.
3. Add the peak rate of runoff from the spillways to the peak rate of runoff from the area below the impoundments.

This method assumes that the peak runoffs from the impoundments and from the area below them reach the discharge point at the same time. This seldom occurs, particularly if the impoundments are in the upper part of the watershed. Consequently the peak rate of runoff obtained by this procedure is usually somewhat higher than the actual peak.

Graded terraces. If the entire watershed is terraced, the storage factor P can be selected from Table 5.4. However, if only part is terraced, the factor should be increased. For example, if 80 acres of a 100-acre watershed are terraced and the terraces average 1,000 feet long, the storage factor would be 0.96.

Runoff Frequency Factor, F. The values given in Table 5.1 are for the peak rate of runoff to be expected in an average ten-year period. If the peak rate of runoff for a different design frequency is desired, the value from the table is multiplied by the appropriate factor F as follows: for the peak rate of runoff expected in an average half-year period, $F = 0.2$;

TABLE 5.4. Storage Factors for Graded Terraces P

Acres in Watershed	Average Length of Terraces (ft)		
	500	1,000	1,500
0–10	0.95	0.90	0.80
11–40	0.97	0.93	0.85
41–100	0.98	0.95	0.90
101–200	0.99	0.97	0.95
201–400	1.00	0.98	0.96
401–600	1.00	0.99	0.97

1 year, $F = 0.3$; 2 years, $F = 0.5$; 5 years, $F = 0.8$; 10 years, $F = 1.0$; 25 years, $F = 1.3$; and 50 years, $F = 1.5$.

Combination of Factors. For watersheds having a number of factors different from those specified in Table 5.1, the peak rate of runoff is determined by multiplying the value from the table by a succession of factors applicable to the watershed being considered.

Example: Estimate the peak rate of runoff to be expected in a twenty-five-year period from a 120-acre watershed located in the southeast corner of Nebraska, and having the following characteristics: (1) Surface soil is moderately coarse textured with a well-drained subsoil. (2) Land slopes average 10 percent. (3) The runoff must travel a maximum distance of 4,700 feet in reaching the discharge point. (4) Eighty acres of row-crop land in the upper half of the watershed are terraced. The terraces average 1,300 feet in length. The crop rows are planted parallel to the terraces. (5) Forty acres of good quality pasture in the lower half of the watershed are not terraced. (6) There are no impoundments except terraces on the watershed.

Procedure: The appropriate factor for each term in the prediction equation is determined from the given conditions: location factor $L = 1.03$; soil factor (above average infiltration) $I = 0.9$; land slope factor $T = 1.07$; shape of watershed factor $S = 0.90$; vegetative cover factor (80 acres of row crops—1.0, and 40 acres of pasture—0.6) $V = 0.87$; contour factor (two-thirds of watershed is terraced) $C = 0.99$; surface storage factor $P = 0.97$; and runoff frequency factor (25 years) $F = 1.3$.

The peak rate of runoff (Q_T) for a 120-acre watershed from Table 5.1 is 285 cfs. Substituting in the equation—

$$Q = Q_T \times L \times I \times T \times S \times V \times C \times P \times F$$
$$Q = 285 \times 1.03 \times 0.9 \times 1.07 \times 0.90 \times 0.87 \times 0.99 \times 0.97 \times 1.3$$
$$Q = 277 \text{ cfs}$$

UNIFORM RATES

In the design of water management systems it is sometimes necessary to determine the uniform rate that will remove a certain depth of runoff from a given area in a specified time. For example, what uniform rate of runoff will remove 2 inches of water from 10 acres in forty-eight hours? A convenient rule of thumb (which is sufficiently accurate for most situations) to use in making this determination is

1 acre-inch per hour $= 1$ cfs

In the above example, we were to remove 20 acre-inches of water in 48 hours, or $20/48 = 0.42$ acre-inch per hour, so the uniform rate of runoff would be 0.42 cfs.

The same rule of thumb can be used to estimate the uniform rate of application necessary to add a certain depth of water to an area in a specified time. For example, if 2 inches of water were applied to 40 acres in 100 hours by an irrigation pump, the rate of application would be $2 \times 40/100 = 0.8$ cfs.

Pumping rates are usually specified in gallons per minute instead of cfs. One cfs is approximately equal to 450 gallons per minute. In the above example, the rate of application would be $0.8 \times 450 = 360$ gallons per minute.

AMOUNT OF RUNOFF

A procedure to follow in estimating the total amount of runoff to be expected is given in Chapter 9.

FLOOD DAMAGES

Flood damages in the United States continue to increase each year in spite of large expenditures to reduce them. These damages already exceed one billion dollars per year and it is estimated that they are increasing by at least 2.7 percent per year (10). Of even greater importance is the occasional loss of life as a result of floods.

Floods are caused by a combination of factors. The great floods on the major rivers are caused by general rains over the watershed, followed by rains of high intensity over certain areas. The general rains fill ponds and lakes and saturate the soil, so that a large percentage of the water from high-intensity rains runs off and moves quickly to flood the tributaries and finally the valley of the mainstream.

The great flood on the Kansas and Missouri rivers in July 1951 was caused in large part by conditions of this kind. Following is an excerpt from a U.S. Weather Bureau report of the conditions causing the flood (11):

By early June, rain and wet soil was hampering farm work in Kansas and this month proved to be the wettest June in 65 years. There were 12 to 18 days of heavy rains over much of the state. By the Fourth of July the situation was serious. Corn and wheat fields were choked with weeds and many small streams were overflowing their banks. On that crucial morning (July 10) when the heavens opened up and unloaded disaster, the soil of Kansas, from the prairie sod of the west to the farmland of the east, was one vast sodden sponge. It would absorb no more. In the next 24 hours several points in the Kansas River Basin were drowned under 8 inches of rain. Thus was the stage set for the most destructive flood in our nation's history.

Floods in headwater streams are caused by local rains of high intensity and of relatively short duration that cover relatively small areas. Flooding may occur on one or more minor tributaries from such rains, but they usually have little effect on the flow in the main stream. These high-intensity rains which produce flash floods on the small streams usually occur during the growing season when flooding of crops in the bottomland is most destructive. The floods remove soil from sloping hillsides and deposit it on the flatlands below, often covering good soil with less productive material. They wash over upland crops and flood those on the lowlands, fill drainage ditches, and destroy fences and roads. Some damage can be repaired, but in most cases these small valleys never completely recover.

Floods can be expected to occur at different times of the year depending on the size of the watershed. The information in Table 5.5 is taken from a study of the seasonal occurrence of floods in the Ohio River Basin in relation to watershed size (9). This distribution varies from region to region, but in most areas the flooding on small watersheds occurs during the growing season when land treatment measures can be effective in increasing infiltration and reducing runoff.

REDUCTION

Flood damages cannot be eliminated but they can be reduced by the following methods:

1. *Reduce peak runoff rates by—*
   a. *Increasing infiltration and moisture-holding capacity of the soil.*
   Land treatment measures which result in a deep, fertile topsoil, a high level of organic matter, good tilth, and good vegetative cover increase the infiltration rate and moisture-holding capacity of the soil, thus reducing runoff and making more water available for crop production. Their effectiveness varies widely depending on a combination of watershed and climatic variables. In some cases on very permeable soils with dense vegetative growth in areas not subjected to intense rainfall, surface runoff from small areas may be eliminated entirely. However there may be runoff from larger areas as a result of subsurface flow. In areas where flood runoff is a problem, land treatment measures which increase the rate of in-

TABLE 5.5. Seasonal Occurrence of Floods in the Ohio River Basin

Drainage Area	Maximum Annual Floods	
	May–Sept.	Oct.–Apr.
(sq mi)	*(% of time)*	
1	99	1
10	87	13
100	66	34
1,000	26	74
10,000	10	90
100,000	5	95

Source: L. L. Harrold, Behavior of water on the land.
J. Soil Water Conserv., vol. 10, no. 6, November 1955.

filtration and the capacity of a soil to store water are effective in reducing small-watershed flooding which results from short, intense rains that occur during the growing season. During this period the soil moisture content is usually low enough so that there is space available in the soil for the additional water. During major floods, however, the soil is usually saturated and land treatment measures have little effect on infiltration rates or peak rates of runoff.

b. *Retarding runoff.* Terraces and diversion channels in most cases retard the runoff and increase the time of concentration for the watershed. The amount of increase depends on the length of the terraces and their direction of flow. Farming operations on terraced watersheds are more nearly on the contour and provide more detention storage. Also, an appreciable amount of detention storage is provided in the terrace channels. The effect of terraces or diversions on peak rates of runoff decreases as the size of the watershed increases. Fifteen to twenty minutes is the maximum increase in the time of concentration that can usually be expected from terraces. This would amount to an appreciable increase on small watersheds but would be an insignificant increase on large watersheds.

Land management practices which result in increased infiltration and retardation of flow are effective in reducing peak rates of runoff from small watersheds and short duration storms, but their effectiveness diminishes as the size of the watershed and the length of the storm increase.

c. *Increasing storage.* Water stored in surface depressions and terraces effectively reduces peak rates of runoff on both small and large watersheds. Level terraces are usually designed to hold up to 2 inches of runoff from the drainage area. These terraces eliminate surface runoff for most rains and effectively reduce the surface runoff from extreme storms; however, subsurface flow from larger watersheds is increased.

Water in detention structures is impounded and discharged automatically over a longer period of time. Another type of storage is provided by flood storage reservoirs in which the outflow is controlled by adjustable gates. The effectiveness of these structures in reducing peak rates of runoff is determined by the percentage of the drainage area that flows into a structure, the volume of storage provided, and the rate at which the water is discharged.

In Figure 5.10(a) the structures on the tributary streams are effective in reducing flood flows immediately downstream.[1] Their

---

1. Definitions of symbols used in figures throughout this book are given in Figure 15.5.

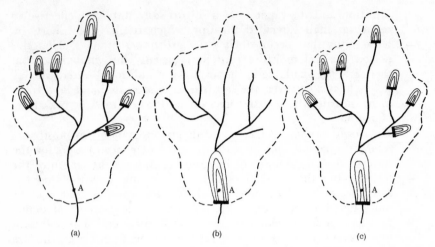

(a)   (b)   (c)

Figure 5.10. Hypothetical watersheds showing some possibilities for location of storage structures.

effectiveness decreases as the distance downstream increases. At some point A, the upstream reservoirs cease to provide adequate flood control in the stream below. A large reservoir of adequate capacity located at A (Fig. 5.10[b]) would be effective in reducing flood flows below A but would have little effect on reducing flooding in the tributary streams above the reservoir. It might possibly create a restriction of flow and cause an increase in flooding of the tributaries immediately upstream from the reservoir. A combination of large and small reservoirs (Fig. 5.10[c]) would be required to reduce flood flows effectively on all streams in the watershed and on the stream below.

In the selection of reservoir sites the value of the land to be inundated by the reservoir must be compared to the value of the land protected from flooding. Large reservoirs are usually of greater depth so the area inundated for a given volume of water impounded is less than for small reservoirs. The initial cost per unit of water stored usually decreases as the size of the reservoir increases. It is important that soil erosion be controlled on the drainage area to prevent sediment and erosion debris from filling the reservoir and reducing its effectiveness.

2. *Control erosion and sedimentation.* The greatest value of land treatment and erosion control practices in reducing flood damage is in their reduction of sedimentation problems downstream. If erosion is not controlled, the eroded soil fills reservoirs and stream channels, reducing their capacity. The sediment in transport in the stream occupies space which reduces the volume of water which can be carried and thus increases flooding. Sediment in the water causes

greater damage when flooding does occur. The deposition of eroded soil on lower slopes and river bottoms usually reduces their productivity.

3. *Increase channel capacity.* This can be accomplished by enlarging the channel, clearing it of vegetation and debris, and straightening it, or by constructing levees to confine the flow.

4. *Control floodplain use.* One reason that flood damages continue to increase is that more intensive use is being made of the floodplain, thus increasing the value of the land as well as the damage that results when flooding occurs. Restricting the floodplains that are subject to flooding to less intensive use is another means of reducing flood damage.

5. *Forecast floods.* A widespread system of flood forecasting would enable people to prepare for floods.

6. *Combination of control methods.* The greatest reduction in flood damage will result from a combination of all methods of control. Land treatment measures are needed to conserve our soil and water resources and reduce flooding on small streams. Reservoirs and channel improvements are needed to reduce flooding on large streams and to protect the life and property of people affected by great floods. Restriction of floodplain use and flood forecasting reduces the damage that results when floods occur.

PROBLEMS

5.1. Select a crop rotation commonly used in your area. At what periods in this rotation would the land be particularly vulnerable to erosion? What is the possibility of intense rains occurring during these periods?

5.2. Use Figure 5.3 to determine the maximum amount and intensity of rainfall to be expected in central Missouri from—
   a. A rain of five minutes duration in a one-hundred-year period.
   b. A rain of two hours duration in a fifty-year period.

5.3 Use Figures 5.2 and 5.3 to determine the maximum amount and intensity of rainfall to be expected in—
   a. Northeastern Ohio from a rain of fifteen minutes duration in a twenty-five-year period.
   b. Central North Dakota from a rain of one hour duration in a one-hundred-year period.
   c. Northwestern Florida from a rain of five minutes duration in a two-year period.

5.4. What effect does soil erosion have on water infiltration rates? Does this vary with different types of soil?

5.5. On a graph with infiltration rate as the ordinate and time as the abscissa, plot the infiltration curves that you would expect as the result of a 3 inch per hour rain lasting for two hours on the following soils:
   a. A uniform sandy soil to a depth of 6 feet.

b. A uniform silt loam soil to a depth of 6 feet.

c. A uniform clay soil to a depth of 6 feet.

d. A silt loam soil underlain at a 12-inch depth by a dense claypan.

e. A soil that contains a large amount of colloidal clay that is extremely dry at the beginning of the rain.

5.6. Plot hydrographs of runoff that you would expect from the 6.5 inch per hour rain if it had lasted for twenty minutes, on the 30-acre watershed and on the 100-acre watershed in Figure 5.4. How often would you expect a rain of this intensity and duration to occur in central Missouri?

5.7. a. Determine the peak rate of runoff in cfs to be expected from a 20-acre watershed under the conditions specified in Table 5.1.

b. This 20-acre watershed will have gradient terraces averaging 1,000 feet long. The terraces will increase to 1,900 feet the distance the runoff must travel in reaching the discharge point. Crop rows will be planted on the contour between terraces. Compute the peak rate of runoff from the terraced watershed in cfs. By what percentage will terracing and contouring reduce the peak rate of runoff from this watershed?

c. Determine the peak rate of runoff to be expected from a 600-acre watershed under the conditions specified in Table 5.1.

d. This 600-acre watershed will have gradient terraces averaging 1,200 feet

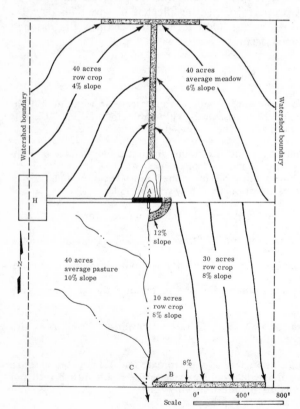

Figure P5.1. A 160-acre watershed to be used in problem solutions.

long. The terraces will increase to 10,400 feet the distance the runoff must travel in reaching the discharge point. Crop rows will be planted on the contour between terraces. Compute the peak rate of runoff from the terraced watershed in cfs. By what percentage will terracing and contouring reduce the peak rate of runoff from this watershed?

e. Explain why the percent reduction in peak rate of runoff by terracing and contouring was less on the larger watershed.

5.8. The watershed shown in Figure P5.1 is located in the southeast corner of Kansas. The soil is a silt loam with a dense subsoil within 12 inches of the surface.

a. Estimate the ten-year, one-year, and twenty-five-year frequency runoff rates from the 80-acre terraced watershed at point *A*, assuming that the reservoir has not been constructed.

b. Estimate the ten-year, two-year, and one-half-year frequency runoff rates from the 30-acre terraced watershed at point *B*.

c. Estimate the twenty-five-year frequency runoff rate from the 160-acre watershed at point *C*, (1) assuming that the reservoir at *A* has not been constructed and (2) assuming that the reservoir at *A* has been constructed and that the peak rate of discharge from the spillways during the twenty-five-year frequency storm will be 50 cfs.

5.9. Three inches of water are to be removed from 40 acres in twenty-four hours. Compute the uniform rate of removal in cfs.

REFERENCES

RAINFALL

1. Carter, V. G. *Man on the landscape*. Washington, D.C.: National Wildlife Federation, 1949.
2. Hershfield, D. M. Rainfall frequency atlas of the United States. U.S. Weather Bureau Tech. Paper 40, May 1961.

RUNOFF

3. Chow, Ven Te. Hydrologic determination of waterway areas for the design of drainage structures for small drainage areas. Univ. Ill. Eng. Expt. Sta. Bull. 462, 1962.
4. ———. *Handbook of applied hydrology*. New York: McGraw-Hill, 1964.
5. Hale, D. D., and Beasley, R. P. Hydrologic investigations of the Burge Branch watershed. Univ. Mo. Agr. Expt. Sta. Res. Bull. 863, May 1964.
6. Langbein, W. B. Annual runoff in the United States. U.S. Geol. Surv. Circ. 52, 1949.
7. U.S. Soil Conservation Service, USDA. Hydrology. National engineering handbook, Sec. 4, 1964.
8. USDA. *Water*. Yearbook of Agriculture, 1955.

FLOOD DAMAGES

9. Harrold, L. L. Behavior of water on the land. *J. Soil Water Conserv.*, vol. 10, no. 6, November 1955.
10. Sewell, W. R. D. Adjustment to floods in the United States. Proc. Summer Inst. Water Resources, vol. 4. Utah State Univ. Civil Eng. Dept., Logan, 1966.
11. U.S. Weather Bureau. Daily Weather Map. July 26, 1951.

6 ~ Grassed Waterways and Underground Outlets

When terracing was first introduced, the runoff from the terraces was discharged into existing draws, fence rows, or road ditches. Gullies soon developed, extending branches back into each terrace channel. This resulted in excessive soil loss and created problems in farming the land.

IN AN ATTEMPT to prevent gullying by concentrated runoff, farmers constructed stabilization structures at intervals down the slope. This system was expensive and difficult to maintain, and the channel between structures was unsightly. Underground tile lines were used to some extent, which eliminated the surface channel, but the expense of installation limited their use in these older systems.

Permanent pastures have sometimes been used as terrace outlets, but in many cases the runoff from the terraces has caused serious gullying. Some pastures having favorable slope conditions have been used successfully as outlets if they are fertilized to maintain an excellent cover and if grazing is controlled to prevent overgrazing and damage to the area when the ground is wet. However, since special care is required in that area of pasture used as an outlet, it is usually found more desirable to set aside a strip of land to serve as a waterway.

GRASSED WATERWAYS

The channel of the waterway has usually been shaped or graded and a vegetative cover established. These channels, called grassed waterways, have been used extensively as outlets for terraces and for the disposal of runoff from diversion channels, stabilization structures, contoured rows, and natural depressions. If a waterway is used as an outlet for terraces, it is commonly called a "terrace outlet."

SHAPES

The cross-sectional shape of the waterway may be trapezoidal, parabolic, or triangular (V-shaped). The trapezoidal and parabolic shapes are most widely used. (See Fig. 6.1 and 6.2.) The parabolic shape is commonly used for waterways located in natural depressions, and the trapezoidal shape for terrace outlets located on ridges or side slopes.

In a parabolic waterway the low flows are concentrated in the center of the waterway, and at high flows the greatest depth and the highest velocity occur near the center also. This is an advantage when the waterway is constructed on flatter land slopes in that it restricts meandering of low flows and reduces possibilities for sediment buildup. The fact that the bottom is sloped to the center results in the terrace water that discharges at the edge of the waterway being spread more rapidly over the outlet rather than being concentrated along the edge.

In a trapezoidal waterway the flow is distributed evenly over the full width. This is an advantage when the waterway is constructed on steeper land slopes in that it reduces the possibility of erosion in the center. Care should be exercised during construction to assure that the bottom of the waterway is level.

LOCATION

A water management plan for the entire farm or area should be developed before a waterway for a specific area is constructed. Factors

Figure 6.1. Cross section of a trapezoidal-shaped waterway showing permanent side dikes.

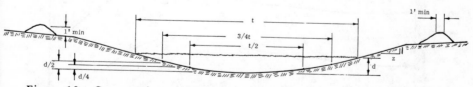

Figure 6.2. Cross section of a parabolic-shaped waterway showing temporary side dikes.

to consider in developing such a plan are discussed in Chapter 11. Waterways should be located so that they will not interfere unduly with farming operations and should not be used as roadways or livestock lanes. Waterways to carry runoff from contoured fields are usually constructed in natural depressions. Those to carry the runoff from terraces may or may not be so located. The location selected should result in maximum farmability of the terraced area and a reasonable cost for waterway construction.

To provide for maximum farmability of the terraced area, convenient access to the areas should be provided, the crop rows should be as long as possible, point rows that require turning of equipment within the field should be kept to a minimum, and the curvature of the terraces and the crop rows planted parallel to the terraces should be minimized. When large machinery is being used, these practices shorten the time required in farming and reduce damage to the soil and crops caused by turning in the field.

Convenient Access. Convenient access can be provided if the outlet is located so that the field entrance is at the upper ends of the terraces. In Figure 6.3 the outlet is in a desirable location if the entrance to the area is at *A*. Any part of the area is accessible by crossing the terraces at the upper end near the fence where they are carrying a minimum amount of water and can be made quite small. If the entrance to the area in Figure 6.3 is at *B*, access to the terraced areas will not be as convenient. In order to enter the areas at the upper end of the terraces, it will be necessary to drive the full length of the field. This will be time

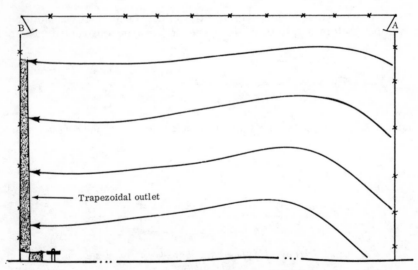

Figure 6.3.   Outlet properly located for access if field entrance is at A.

consuming and probably will not be done. The tendency will be to use the outlet as a roadway; or if the outlet is fenced, the terraces will be crossed near the lower end where they carry the maximum amount of water and are quite large. Use of the outlet as a roadway should not be permitted since it will destroy the grass and cause excessive erosion. Crossing the terraces at the lower end will reduce the terrace height and result in overtopping and will cause ponding in the terrace channel, resulting in miring down and breakage of machinery.

If the entrance to the area is at *B,* the outlet location shown in Figure 6.4 will provide convenient access except for the small triangular area at the far end of the field. The outlet must be crossed to farm this area which is too small and has rows too short to be farmed as a separate field. An alternate outlet is shown by the dotted lines. This would be more difficult to construct because it is located on a side hill and large quantities of earth would have to be moved to make it level. This would be expensive and would result in less desirable conditions for the establishment and maintenance of grass.

When the position of existing fences results in areas that are diffi-cult to farm, rearrangement should be considered in developing the water management plan. For example, with the outlet located as in Figure 6.4, it would be desirable to remove the existing fence so that the area to the right of the outlet could be included in another field.

Terrace Curvature and Row Length. The optimum length of crop rows and the curvatures of the terraces will influence the location of the waterway. The topography in Figure 6.5 is quite irregular. With the

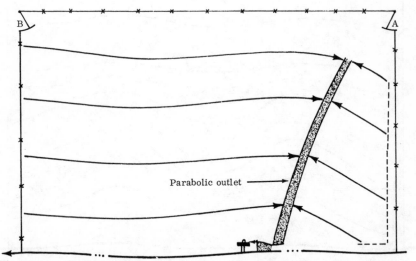

Figure 6.4. Outlet properly located if the field entrance is at B. *The dotted line indicates an alternate outlet location.*

outlet located as in this sketch, the terraces are long and some of the crop rows extend the full length of the area. However, the terraces and the crop rows parallel to the terraces are quite crooked. Also, since it is not feasible to construct the terraces parallel, numerous point rows exist, which necessitate additional turning in the field when farming.

If the outlets are located in the natural depressions as in Figure 6.6, the terraces can be made less crooked and more nearly parallel. However, the outlets themselves interfere with the farming operations. In order to farm the full length of the area, it is necessary to cross the outlets. Additional time is required to raise the equipment while crossing, and maintaining an adequate stand of grass will be difficult. If the outlets are not crossed, the area must be farmed as three separate fields. The rows will be shorter, requiring more turning time, and additional land will be taken up in outlets and turn-row areas.

If the grassed outlets in Figure 6.6 were replaced by underground outlets, the farmability of the area would be improved. Good judgment is required in selecting the type of outlet and the location that will result in maximum farmability on any given area.

<h2 style="text-align:center">DESIGN</h2>

A waterway should normally be designed for the peak rate of runoff expected in a ten-year period. However, on flat slopes where occasional overflow would not cause significant damage and it is difficult to obtain adequate depth, a lower peak rate of runoff may be used. The waterway

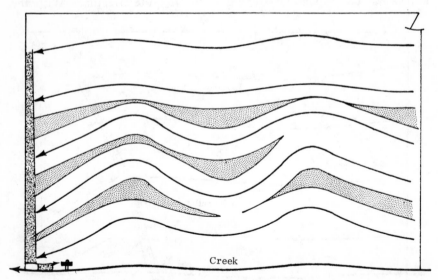

Creek

Figure 6.5.   Terrace location and point row areas when the outlet is located on the edge of the area.

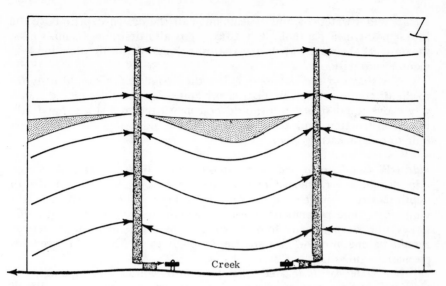

Figure 6.6. Terrace location and point row areas when the outlets are located in the natural depressions.

should be of such width that the resulting velocity will not erode the channel when the grass is short and of such depth that it will carry the peak rate of runoff when the grass is long.

Width. The width of waterway determined by the design procedure (given in the Formulas and Procedures section which follows) is designated the *design width*. This width will result in a velocity in the waterway equal to the maximum permissible velocity when the grass is short. The waterway should not be constructed narrower than the design width, or the velocity will be excessive. However, the waterway can be made wider than the design width if it will facilitate the construction or maintenance of the waterway. If a waterway greater than 60 feet wide is required, division into smaller waterways should be considered. If water enters from both sides of a wide waterway, a small dike approximately 0.4 foot high should be constructed in the center to divide the flow. If the waterway is to be crossed during farming operations, a minimum top width of 20 feet is desirable.

Depth. The depth of flow determined by the design procedure which follows is designated the *design depth*. If the waterway is used as a terrace outlet, it is usually necessary to construct it deeper than the design depth in order that the terraces can be cut into the waterway on proper alignment and with adequate grade. The bottom of the terrace channel at the waterway should usually be about 0.2 foot above the bottom of a trapezoidal waterway and at least one-half the design depth above the lowest point in a parabolic waterway. It is also desirable in most cases to construct the waterway deeper than the design depth to allow for sediment accumulation.

The shape of the waterway below the design depth should conform to the dimensions given in Figures 6.1 and 6.2. The sides of the waterway above the design depth should be constructed to a slope that facilitates construction and maintenance. These side slopes are normally four to one or flatter.

Side Dikes and Freeboard. If permanent side dikes are needed to contain the flow, a freeboard of 0.3 foot should be added to the design depth to assure future capacity, since the dikes will be reduced in height with time. The permanent side dikes should be constructed with side slopes not steeper than four to one. When possible, the waterway should be cut into the ground deep enough to eliminate the need for permanent side dikes.

If a waterway is located where overflow would cause damage, the freeboard deemed necessary to provide the desired protection should be added.

Other Considerations. If there is a variation in the slope of the land on which the waterway is to be constructed or a variation in the peak rate of runoff, it is advisable to determine the size of the waterway at several points. Some additional factors must be considered in design if waterways are used as terrace outlets.

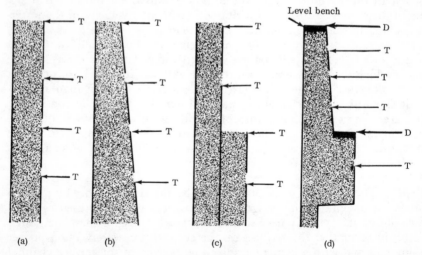

Figure 6.7. Variation in the shape of terrace outlets. (T = terrace, D = diversion.)

Figure 6.8. Parallel outlets on different levels.

If the land slope on which the outlet is constructed is quite uniform, a wider outlet is required at the lower end where it is carrying the greatest amount of runoff. This outlet may be made of uniform width equal to the width at the lower end (Fig. 6.7[a]), or it may be tapered (Fig. 6.7[b]). The uniform width is easier to maintain, but more land is taken up by the outlet; whereas the tapered outlet requires less land and also results in a more uniform distribution of water, which is quite important in a wide trapezoidal outlet. The amount of land in the outlet, the cost of construction, the ease of maintenance, and the distribution of water over the outlet must be evaluated in selecting the shape to be constructed at a given location.

Where a width greater than 60 feet is required, it may be more desirable to divide the flow into two smaller outlets parallel to each other (Fig. 6.7[c]). Parallel outlets may also be used to advantage where a wide trapezoidal outlet is to be constructed on land that has considerable side slope. The two outlets can be built on different levels, requiring less earth movement (Fig. 6.8).

Locations exist where an outlet will be built on sloping land and also across bottomland with relatively flat slopes. A narrower outlet can be constructed across the bottomland if the extra land gained for cultivation justifies the inconvenience of maintaining the irregular shape (Fig. 6.7[d]). On irregular land, the size of outlet required may increase or decrease in width several times down the slope. In order to improve construction and maintenance and prevent concentration of water along one edge, it is necessary at some locations to construct the outlet wider than the design width, but it should never be made narrower.

If a diversion channel carrying a large volume of water enters a trapezoidal-shaped outlet, the water from the diversion should be spread over the required width of outlet as soon as possible to prevent erosion. This can usually be accomplished by making a level bench at the point where the diversion channel enters the outlet (Fig. 6.7[d]).

Formulas and Procedures. The size of grassed waterways and other channels such as terraces, diversions, cross-slope channels, and drainage ditches, used to convey water, can be determined by the following formulas and procedures. As noted earlier, the shape of the constructed channels may be trapezoidal, parabolic, or triangular. The geometric characteristics of trapezoidal channels with four-to-one side slopes are given in Figure 6.9; and with side slopes other than four to one in Table 6.1.

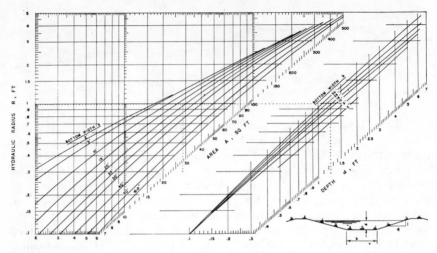

Figure 6.9. Dimensions of trapezoidal channels with four-to-one side slopes. Reproduced from U.S. Soil Conservation Service, USDA, Handbook of channel design for soil and water conservation, SCS-TP-61, June 1954.

The geometric characteristics of parabolic channels are given in Figure 6.10 and of triangular channels in Figure 6.11.

Velocity. Open channels must be designed to carry the expected runoff at a suitable velocity. As the velocity in a channel increases, the size of channel required to carry a given flow decreases. The carrying capacity of a channel can be computed by the formula $Q = AV$,

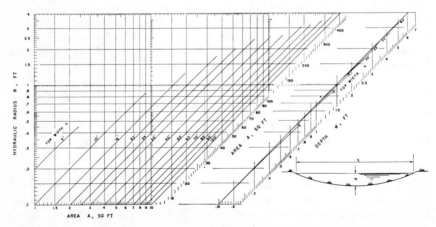

Figure 6.10. Dimensions of parabolic channels. See Figure 6.2 for additional dimensions. Reproduced from U.S. Soil Conservation Service, USDA, Handbook of channel design for soil and water conservation, SCS-TP-61, June 1954.

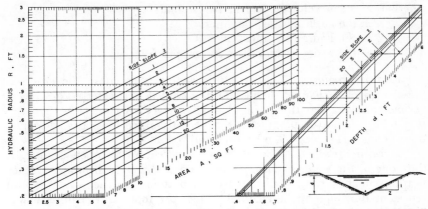

Figure 6.11. Dimensions of triangular channels. Reproduced from U.S. Soil Conservation Service, USDA, Handbook of channel design for soil and water conservation, SCS-TP-61, June 1954.

where Q = capacity, in cfs

A = cross-sectional area of the channel, in square feet

V = the average velocity, in feet per second

The maximum permissible velocity is usually limited by the erodibility of the soil in which the channel is constructed and the type and condition of the vegetation or other material lining the channel. The velocity should be kept low enough so that excessive erosion of the channel will not occur. The minimum permissible velocity should be high enough to avoid excessive deposition.

The velocity in an open channel may be computed by Manning's formula:

$$V = \frac{1.49}{n} R^{2/3} S^{1/2}$$

TABLE 6.1. Geometric Characteristics of Trapezoidal Channels

Side Slopes	A	wp^a
1:1	$db + d^2$	$b + 2.82e$
2:1	$db + 2d^2$	$b + 2.24e$
3:1	$db + 3d^2$	$b + 2.11e$
4:1	$db + 4d^2$	$b + 2.06e$
5:1	$db + 5d^2$	$b + 2.04e$
6:1	$db + 6d^2$	$b + 2.03e$
7:1	$db + 7d^2$	$b + 2.02e$
8:1	$db + 8d^2$	$b + 2.02e$
9:1	$db + 9d^2$	$b + 2.01e$
10:1	$db + 10d^2$	$b + 2.01e$

[a] Accurate to the last digit indicated.

Note: $R =$ Area/wp, and $t = b + 2e$. See Figure 6.12 for other symbols and dimensions.

where V = mean velocity, in feet per second
$\quad n$ = the coefficient of roughness
$\quad R$ = the hydraulic radius, in feet
$\quad S$ = the hydraulic gradient, in feet per foot

Hydraulic radius. The hydraulic radius is equal to the cross-sectional area of a channel divided by the wetted perimeter. The wetted perimeter is the length of the channel surface in contact with water (Fig. 6.12). The hydraulic radius can be determined by charts and equations given in Figures 6.9, 6.10, 6.11 and Table 6.1.

Hyraulic gradient. The hydraulic gradient in an open channel is the slope of the water surface. In most open channels the water surface is approximately parallel to the channel bottom, in which case S would represent the slope of the channel in feet per foot. If a channel had a gradient or slope of 0.3 percent, then S would equal 0.003 foot per foot.

Coefficient of roughness. In applying Manning's formula to the design of open channels, the greatest difficulty lies in the selection of the coefficient of roughness. To select a value of n, the designer must estimate the resistance to flow in a given channel. Table 6.2 gives values of n for channels of different types. Consideration of these values along with an exercise of good judgment in evaluating the factors affecting n enables the designer to make a reasonable decision as to the value of n to use in design. The following factors affect the value of n:

1. Surface roughness. The size and shape of the material forming the sides and bottom of the channel affect n. A channel lined with large, rough rock offers greater retardance to flow than a channel lined with uniform silt particles.
2. Vegetation. Vegetation affects the surface roughness of the channel, but more important, it retards the flow and reduces the capacity. The coefficient of roughness depends upon the type of vegetation and its height and density in relation to the depth of water.

    Extensive tests to determine the effect of vegetation on the roughness coefficient have been conducted at the Missouri Soil Conservation Experiment Farm near McCredie (7), at the Stillwater Outdoor Hydraulic Laboratory at Stillwater, Oklahoma (3, 5), and at an

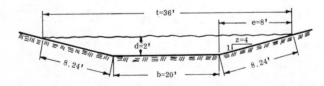

Area=56 sq ft, wetted perimeter=36.48 feet,
hydraulic radius=56/36.48=1.53 feet

Figure 6.12.   Dimensions of a trapezoidal channel with a 20-foot bottom width, 2-foot depth, and four-to-one side slopes.

TABLE 6.2. Values of the Coefficient of Roughness n

Type of Channel	n
Lined channels	
Concrete, float finish	0.015
Concrete, trowel finish	0.013
Wood, planed	0.012
Wood, unplaned	0.013
Brick, in cement mortar	0.015
Excavated earth channels	
Straight and uniform	
Clean	0.022
Bottom clean, low vegetation on sides	0.035
Bottom clean, weeds and low brush on sides	0.050
Underwater growth on bottom, dense growth of brush, and weeds on sides	0.100
Crooked and irregular	
Dense vegetative growth on sides and bottom	0.200
Grassed waterways	
See Figures 6.13, 6.14, 6.15, and 6.16	

outdoor hydraulic laboratory near Spartanburg, South Carolina (6). The following discussion of n is based, in part, on these tests.

Values for the coefficient of roughness n for grassed waterways on a 5 percent slope are given in Figure 6.13. Curve A gives values for a medium length Bermuda grass sod, curve B gives values for a headed bluegrass sod, curve C gives values for a short Bermuda grass sod,

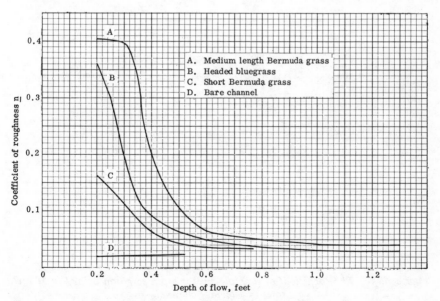

Figure 6.13. The coefficient of roughness n for sod-forming grasses in a grassed waterway on a 5 percent slope.

and curve *D* gives values for a bare soil. The roughness coefficient or flow retardance of grasses was high for shallow depths. As the depth of flow increased, the longer grasses tended to bend over and retardance decreased rapidly. When the grasses were bent over and submerged, the *n* values leveled out and approached constant value. The values for a bare soil were much lower and tended to increase slightly with depth because the channel surface became rougher as the depth of flow and velocity increased.

Values for the coefficient of roughness *n* for a terrace channel with grain sorghum planted in 40-inch rows parallel to the channel is given in Figure 6.14. The grade in the terrace channel was 0.1 percent. The grain sorghum was headed out and the plant height varied from 16 to 54 inches, the average being 43 inches. The coefficient of roughness increased as the depth of flow increased because of the greater density of foliage at the greater depth. The plants were not submerged and the velocity was not great enough to bend the stalks over.

3. Channel irregularity and curvature. In most constructed channels, changes in cross section and curvature are made gradually enough that *n* is not affected an appreciable amount.

4. Channel slope. With the same depth of flow, a grassed waterway on a steeper slope will have a higher velocity. This higher velocity will result in greater bending of the grasses and a lower value of *n*.

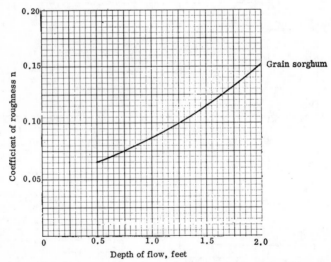

Figure 6.14.   The coefficient of roughness n for grain sorghum drilled in 40-inch rows in a terrace channel.

TABLE 6.3. Maximum Permissible Velocities for Open Channels, Feet per Second

Cover	More Erodible Soils[a]			Less Erodible Soils[b]		
	Percent slope			Percent slope		
	0–5	6–10	over 10	0–5	6–10	over 10
None						
Cultivated	1.5	...	...	2.0	...	...
Not cultivated	2.0	...	...	2.5	...	...
Annual grasses						
Sparse stand	2.5	...	...	3.0	...	...
Good stand	3.0	...	...	3.5	...	...
Average density grasses *(bluegrass, fescue, brome)*						
Fair stand	3.0	2.5	c	4.0	3.0	2.5
Good stand	4.0	3.5	3.0	5.0	4.0	3.5
Excellent stand	5.0	4.5	4.0	6.0	5.0	4.5
Dense sod grasses (Bermuda)						
Excellent stand	6.0	5.0	4.0	7.0	6.0	5.0

[a] More erodible soils are generally those that have a high content of fine sand or silt and lower plasticity. Typical soil textures are fine sand, silt, sandy loam, and silty loam.

[b] Less erodible soils are generally those with a higher clay content and higher plasticity. Typical soil textures are silty clay, sandy clay, and clay.

[c] Must have good quality vegetation on these slopes.

Dimensions. To determine the dimensions of an open channel for a given use, the following procedure is suggested:

1. Compute the rate of runoff Q to be carried by the channel.
2. Determine the slope S in the channel at the point being considered.
3. Determine the type of vegetation, if any, and select the maximum permissible velocity—the greatest mean velocity that will not result in serious erosion of the channel. Table 6.3 gives suggested permissible velocities to use if more specific values are not available.
4. Calculate the required cross-sectional area, $A = Q/V$.
5. Determine the dimensions of the channel that will provide the desired velocity and carrying capacity. Design charts have been prepared which are useful in the design of grassed waterways. A trial-and-error procedure is required for the design of other channels.

Short grass. The waterway cross section is first designed so that excessive velocity and erosion do not occur when the grass in the waterway is short and offers the least resistance to flow. The following steps are suggested:

1. Using the permissible velocity and the slope in the waterway, obtain the value of the required hydraulic radius R.

a. Use Figure 6.15 if a reasonably good stand of grass is to be established and maintained in the channel.

b. Use Figure 6.16 for those channels where it will not be possible to establish a stand of permanent grass, and the vegetation will consist primarily of annual grasses.

2. Using the above value of R and the value of A previously determined, obtain the dimensions of the channel.

a. For trapezoidal channels with four-to-one side slopes use Figure 6.9. For side slopes other than four to one use the information given in Table 6.1 and a trial-and-error procedure to select the channel width and depth that will give the required values of A and R.

b. Obtain the dimensions of a parabolic channel from Figure 6.10.

c. Obtain the dimensions of a triangular channel from Figure 6.11.

Long grass. The waterway should be designed so that it will have adequate capacity when the grass is long and offers maximum retardance.

1. From Table 6.4 obtain the additional depth of flow resulting from a rank growth of grass in the channel.

2. Figure the total top width of the waterway using the additional depth and the appropriate side slope. In figuring the additional width required for parabolic waterways, side slopes of four to one or flatter should be used.

Example: A waterway with a parabolic cross section is to be designed to carry 100 cfs. The slope in the waterway is 5 percent. A good fescue sod is to be established and maintained. The soil is easily eroded.

1. The maximum permissible velocity is 4 feet per second (Table 6.3).

2. The required cross-sectional area is $100/4 = 25$ square feet.

3. The value of R from Figure 6.15 is 0.43.

4. The top width is 58 feet and the depth of flow is 0.64 foot (Fig. 6.10).

This is the width and depth of flow that will result in a velocity of 4 feet per second with short grass in the channel. The width and depth required to carry the 100 cfs with a rank growth of vegetation in the channel should be determined.

1. The additional depth resulting from the rank vegetation is 0.3 foot (Table 6.4).

2. If side slopes of five to one are to be used on this section of the waterway, an additional width of $2(0.3 \times 5) = 3$ feet will be needed.

3. The total top width of the waterway will be $58 + 3 = 61$ feet and the depth of flow with rank vegetation will be $0.64 + 0.3 = 0.94$ foot.

R

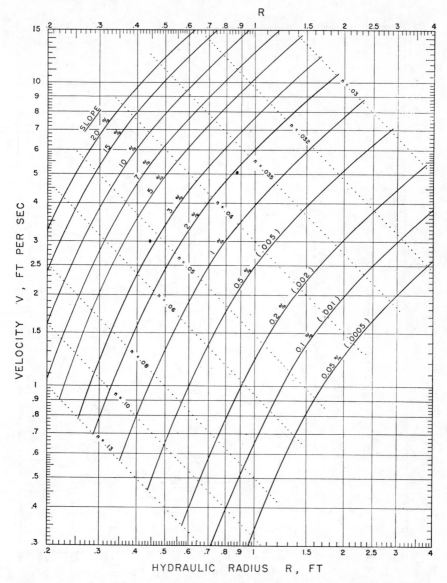

Figure 6.15. Solution of Manning's formula for low vegetal retardance D. Reproduced from U.S. Soil Conservation Service, USDA, Handbook of channel design for soil and water conservation, SCS-TP-61, June 1954.

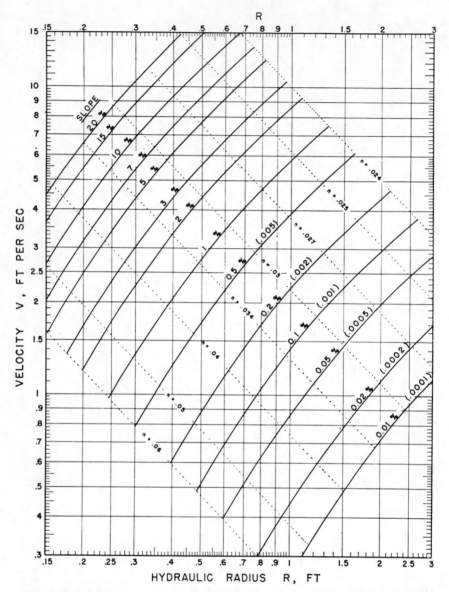

Figure 6.16. Solution of Manning's formula for very low vegetal retardance E. Reproduced from U.S. Soil Conservation Service, USDA, Handbook of channel design for soil and water conservation, SCS-TP-61, June 1954.

TABLE 6.4. Additional Depth of Flow in Grassed Waterways Resulting from a Rank Growth of Vegetation in the Channel

Channel Slope (%)	Additional Depth of Flow (ft)
0–0.9	0.7
1.0–1.9	0.6
2.0–2.9	0.5
3.0–4.9	0.4
5.0 and over	0.3

Channels other than grass. Most channels other than grassed waterways (such as drainage ditches) have either a trapezoidal or a triangular cross section. Charts which can be used to select the required R for a given V (Figs. 6.15 and 6.16) are not available for channels other than grassed waterways, so the following procedure is suggested:

1. Determine Q, S, V, and A as above. (From this point a trial-and-error procedure is required.)
2. Select channel dimensions which will give the correct area and can be assumed to result in the correct velocity.
3. Solve for R using Figure 6.9 or Table 6.1 for trapezoidal channels and Figure 6.11 for triangular channels.
4. Obtain the value of $R^{2/3}$ from Table 6.5.
5. Obtain the value of $S^{1/2}$ from Table 6.6.
6. Select an appropriate value of n from suggestions given in Table 6.2 or Figures 6.13 and 6.14.
7. Substitute values of $R^{2/3}$, $S^{1/2}$, and n in Manning's equation,

$$V = \frac{1.49}{n} R^{2/3} S^{1/2}$$

and solve for V. If this V is equal to the maximum permissible velocity, you have selected the correct channel dimensions. If the computed value of V is greater than the maximum permissible velocity, select a wider channel with less depth; if less than the maximum permissible velocity, select a narrower channel with a greater depth. Repeat steps 3, 4, 5, 6, and 7. Continue this trial-and-error procedure until the channel dimensions are selected which result in the correct area and velocity.

Example: A drainage ditch with a trapezoidal cross section and three-to-one side slopes is to carry 50 cfs. The ditch grade is to be 0.15 percent. It is to be constructed in a more erodible soil. Assume that the ditch will have some water in it most of the time so that the ditch bottom will be clean and there will be low vegetation on the sides.

TABLE 6.5. Values of $R^{2/3}$

Number	.00	.01	.02	.03	.04	.05	.06	.07	.08	.09
.0	.000	.046	.074	.097	.117	.136	.153	.170	.186	.201
.1	.215	.229	.243	.256	.269	.282	.295	.307	.319	.331
.2	.342	.353	.364	.375	.386	.397	.407	.418	.428	.438
.3	.448	.458	.468	.477	.487	.497	.506	.515	.525	.534
.4	.543	.552	.561	.570	.578	.587	.596	.604	.613	.622
.5	.630	.638	.647	.655	.663	.671	.679	.687	.695	.703
.6	.711	.719	.727	.735	.743	.750	.758	.765	.773	.781
.7	.788	.796	.803	.811	.818	.825	.832	.840	.847	.855
.8	.862	.869	.876	.883	.890	.897	.904	.911	.918	.925
.9	.932	.939	.946	.953	.960	.966	.973	.980	.987	.993
1.0	1.000	1.007	1.013	1.020	1.027	1.033	1.040	1.046	1.053	1.059
1.1	1.065	1.072	1.078	1.085	1.091	1.097	1.104	1.110	1.117	1.123
1.2	1.129	1.136	1.142	1.148	1.154	1.160	1.167	1.173	1.179	1.185
1.3	1.191	1.197	1.203	1.209	1.215	1.221	1.227	1.233	1.239	1.245
1.4	1.251	1.257	1.263	1.269	1.275	1.281	1.287	1.293	1.299	1.305
1.5	1.310	1.316	1.322	1.328	1.334	1.339	1.345	1.351	1.357	1.362
1.6	1.368	1.374	1.379	1.385	1.391	1.396	1.402	1.408	1.413	1.419
1.7	1.424	1.430	1.436	1.441	1.447	1.452	1.458	1.463	1.469	1.474
1.8	1.480	1.485	1.491	1.496	1.502	1.507	1.513	1.518	1.523	1.529
1.9	1.534	1.539	1.545	1.550	1.556	1.561	1.566	1.571	1.577	1.582
2.0	1.587	1.593	1.598	1.603	1.608	1.613	1.619	1.624	1.629	1.634
2.1	1.639	1.645	1.650	1.655	1.660	1.665	1.671	1.676	1.681	1.686
2.2	1.691	1.697	1.702	1.707	1.712	1.717	1.722	1.727	1.732	1.737
2.3	1.742	1.747	1.752	1.757	1.762	1.767	1.772	1.777	1.782	1.787
2.4	1.792	1.797	1.802	1.807	1.812	1.817	1.822	1.827	1.832	1.837
2.5	1.842	1.847	1.852	1.857	1.862	1.867	1.871	1.876	1.881	1.886
2.6	1.891	1.896	1.900	1.905	1.910	1.915	1.920	1.925	1.929	1.934
2.7	1.939	1.944	1.949	1.953	1.958	1.963	1.968	1.972	1.977	1.982
2.8	1.987	1.992	1.996	2.001	2.006	2.010	2.015	2.020	2.024	2.029
2.9	2.034	2.038	2.043	2.048	2.052	2.057	2.062	2.066	2.071	2.075
3.0	2.080	2.085	2.089	2.094	2.099	2.103	2.108	2.112	2.117	2.122
3.1	2.126	2.131	2.135	2.140	2.144	2.149	2.153	2.158	2.163	2.167
3.2	2.172	2.176	2.180	2.185	2.190	2.194	2.199	2.203	2.208	2.212
3.3	2.217	2.221	2.226	2.230	2.234	2.239	2.243	2.248	2.252	2.257
3.4	2.261	2.265	2.270	2.274	2.279	2.283	2.288	2.292	2.296	2.301
3.5	2.305	2.310	2.314	2.318	2.323	2.327	2.331	2.336	2.340	2.345
3.6	2.349	2.353	2.358	2.362	2.366	2.371	2.375	2.379	2.384	2.388
3.7	2.392	2.397	2.401	2.405	2.409	2.414	2.418	2.422	2.427	2.431
3.8	2.435	2.439	2.444	2.448	2.452	2.457	2.461	2.465	2.469	2.474
3.9	2.478	2.482	2.486	2.490	2.495	2.499	2.503	2.507	2.511	2.516
4.0	2.520	2.524	2.528	2.532	2.537	2.541	2.545	2.549	2.553	2.558
4.1	2.562	2.566	2.570	2.574	2.579	2.583	2.587	2.591	2.595	2.599
4.2	2.603	2.607	2.611	2.616	2.620	2.624	2.628	2.632	2.636	2.640
4.3	2.644	2.648	2.653	2.657	2.661	2.665	2.669	2.673	2.677	2.681
4.4	2.685	2.689	2.693	2.698	2.702	2.706	2.710	2.714	2.718	2.722
4.5	2.726	2.730	2.734	2.738	2.742	2.746	2.750	2.754	2.758	2.762
4.6	2.766	2.770	2.774	2.778	2.782	2.786	2.790	2.794	2.798	2.802
4.7	2.806	2.810	2.814	2.818	2.822	2.826	2.830	2.834	2.838	2.842
4.8	2.846	2.850	2.854	2.858	2.862	2.865	2.869	2.873	2.877	2.851
4.9	2.885	2.889	2.893	2.897	2.901	2.904	2.908	2.912	2.916	2.920

TABLE 6.6. Values of $S^{1/2}$

Number	.−0	.−1	.−2	.−3	.−4	.−5	.−6	.−7	.−8	.−9
.001	.03162	.03317	.03464	.03606	.03742	.03873	.04000	.04123	.04243	.04359
.002	.04472	.04583	.04690	.04796	.04899	.05000	.05009	.05196	.05292	.05385
.003	.05477	.05568	.05657	.05745	.05831	.05916	.06000	.06083	.06164	.06245
.004	.06325	.06403	.06481	.06557	.06633	.06708	.06782	.06856	.06928	.07000
.005	.07071	.07141	.07211	.07280	.07348	.07416	.07483	.07550	.07616	.07681
.006	.07746	.07810	.07874	.07937	.08000	.08062	.08124	.08185	.08246	.08307
.007	.08367	.08426	.08485	.08544	.08602	.08660	.08718	.08775	.08832	.08888
.008	.08944	.09000	.09055	.09110	.09165	.09220	.09274	.09327	.09381	.09434
.009	.09487	.09539	.09592	.09644	.09695	.09747	.09798	.09849	.09899	.09950
.010	.10000	.10050	.10100	.10149	.10198	.10247	.10296	.10344	.10392	.10440
.01	.1000	.1049	.1095	.1140	.1183	.1225	.1265	.1304	.1342	.1378
.02	.1414	.1449	.1483	.1517	.1549	.1581	.1612	.1643	.1673	.1703
.03	.1732	.1761	.1789	.1817	.1844	.1871	.1897	.1924	.1949	.1975
.04	.2000	.2025	.2049	.2074	.2098	.2121	.2145	.2168	.2191	.2214
.05	.2236	.2258	.2280	.2302	.2324	.2345	.2366	.2387	.2408	.2429
.06	.2449	.2470	.2490	.2510	.2530	.2550	.2569	.2588	.2608	.2627
.07	.2646	.2665	.2683	.2702	.2720	.2739	.2757	.2775	.2793	.2811
.08	.2828	.2846	.2864	.2881	.2898	.2915	.2933	.2950	.2966	.2983
.09	.3000	.3017	.3033	.3050	.3066	.3082	.3098	.3114	.3130	.3146
.10	.3162	.3178	.3194	.3209	.3225	.3240	.3256	.3271	.3286	.3302

1. The maximum permissible velocity to prevent erosion of the clean bottom is 2.0 feet per second. (See Table 6.3.)
2. The required cross-sectional area is $50/2 = 25$ square feet.
3. Select a depth of flow $d = 2.5$ feet.
4. From Table 6.1—
   a. Area $= db + 3d^2$, substitute $A = 25$, $d = 2.5$, and solve for b:

$$25 = 2.5b + 3 \times 2.5^2$$
$$b = 2.5 \text{ feet}$$

   b. $e = 3 \times 2.5 = 7.5$ feet
   c. $wp = 2.5 + 2.11e = 2.5 + 2.11 \times 7.5 = 18.33$ feet
   d. $R = A/wp = 25/18.33 = 1.36$
5. $R^{2/3} = 1.23$ (Table 6.5)
6. $S^{1/2} = 0.0015^{1/2} = 0.0387$ (Table 6.6)
7. $n = 0.035$ (Table 6.2)
8. Substituting in $V = 1.49/n \ R^{2/3} \ S^{1/2}$, $V = 1.49/0.035 \times 1.23 \times 0.0387 = 2.03$ feet per second

This velocity is slightly higher than the permissible velocity of 2.0 feet per second, but the difficulty of constructing the ditch with a 2.5-foot bottom would be of greater importance. A ditch with a shallower depth and a wider bottom would give a lower velocity and be easier to construct.

1. Select a depth of flow of 2.2 feet.
2. From Table 6.1—
   a. Area $= db + 3d^2$ with $A = 25$ and $d = 2.20$
   b. $b = 4.8$ feet
   c. $e = 3 \times 2.2 = 6.6$ feet
   d. $wp = 4.8 + 2.11 \times 6.6 = 18.73$
   e. $R = A/wp = 25/18.73 = 1.33$
3. $R^{2/3} = 1.209$
4. $S^{1/2} = 0.0387$
5. $n = 0.035$
6. Substituting in

$$V = 1.49/n \; R^{2/3} \; S^{1/2}$$
$$V = 1.49/0.035 \times 1.209 \times 0.0387$$
$$V = 1.99 \text{ feet per second}$$

7. $Q = AV = 25 \times 1.99 = 49.9$ cfs. The velocity and capacity computed are close enough to those required for all practical purposes.
8. The ditch has a bottom width of 4.8 feet, a depth of 2.2 feet, and a top width of $4.8 + 2(2.2 \times 3) = 18.0$ feet.

The channel should be designed with sufficient depth to provide adequate capacity if a rank growth of vegetation grows on the sides and an underwater growth of vegetation occurs on the bottom which will result in an n value of 0.100 (Table 6.2). This is a trial-and-error procedure to select the depth of flow which will result in a channel with the desired capacity.

1. Select a depth of flow of 4 feet.
2. Area $= bd + 3d^2 = 4.8 \times 4 + 3 \times 4^2 = 67.2$ square feet
3. $e = 3 \times 4 = 12$
4. $wp = b + 2.11e = 4.8 + 2.11 \times 12 = 30.12$
5. $R = A/wp = 67.2/30.12 = 2.23$
6. $R^{2/3} = 1.707$
7. $S^{1/2} = 0.0387$
8. $V = 1.49/n \; R^{2/3} \; S^{1/2} = 1.49/0.10 \times 1.707 \times 0.0387 = 0.98$ feet per second
9. $Q = 67.2 \times 0.98 = 66$ cfs

This is greater capacity than the 50 cfs required, so a shallower depth can be used.

1. Estimate a depth of 3.6 feet.
2. Area $= bd + 3d^2 = 4.8 \times 3.6 + 3 \times (3.6)^2 = 56.16$
3. $e = 3 \times 3.6 = 10.8$
4. $wp = b + 2.11e = 4.8 + 2.11 \times 10.8 = 27.6$
5. $R = A/wp = 56.16/27.6 = 2.03$
6. $R^{2/3} = 1.603$
7. $S^{1/2} = 0.0387$
8. $V = 1.49/n\ R^{2/3}\ S^{1/2} = 1.49/0.10 \times 1.603 \times 0.0387 = 0.924$
9. $Q = AV = 56.16 \times 0.924 = 51.9$ cfs which is close enough to the required Q for all practical purposes.
10. This ditch will have a bottom width of 4.8 feet, a depth of 3.6 feet, and a top width of $4.8 + 2(3.6 \times 3) = 26.4$ feet.

Note the much larger ditch required when vegetative growth is allowed to grow up and restrict the flow. This emphasizes the importance of adequate maintenance of open channels.

CONSTRUCTION

If a waterway is to be used as a terrace outlet, it should be constructed one year ahead so that a dense sod will be established by the time the terraces are built. It should also be constructed deep enough so that terraces can be brought in on proper alignment with adequate grade.

The waterway should be staked out to the size required and built to the proper width and depth. Small waterways may be constructed with farm equipment. Motor graders, bulldozers, or tractor-drawn scrapers are commonly used on large waterways.

Temporary side dikes should be constructed to prevent water from the surrounding area from running into the waterway. They should be removed when the vegetation is well established, and the soil spread so that it will not interfere with the proper functioning of the waterway. Permanent side dikes may be needed for waterway capacity; if so, they should be constructed with side slopes no steeper than four to one.

After the rough grading has been completed, fertilizer, lime, and manure, as indicated by soil test, should be applied to the waterway and plowed or worked deeply into the soil. The waterway should then be releveled or reshaped and a firm seedbed prepared.

At the proper time the waterway should be seeded to grasses that are sod forming and well adapted to the soil, the site, and the locality. Commonly used grasses include tall fescue, smooth brome, Kentucky bluegrass, redtop, Bermuda grass, and reed canary grass. A seed of good quality should be used and seed applied at a rate approximately three times as heavy as normally required for a meadow seeding.

It is important to prevent erosion in the waterway while the grass is becoming established. Where possible, small furrows should be placed across the waterway at approximately every terrace spacing to collect and divert the runoff before it accumulates in sufficient volume to cause damage. When the grass is 3 to 4 inches high, these furrows should be leveled out and the disturbed area reseeded. Where it is not possible to divert the runoff from the waterway, as would be the case for a waterway in a deep depression, quick-growing cover crops and mulches of hay, straw, or other material should be used to give protection from erosion until the grass is established.

MAINTENANCE

After grassed waterways are properly designed and constructed and sod is established, their success depends upon proper maintenance. Turning or driving tractors and farm machinery in the outlet should not be permitted since they leave depressions or deposits of soil which result in concentrations of water and formation of gullies. Overgrazing by livestock or grazing when the waterway is wet should not be permitted. Weeds should be controlled by spraying or mowing. Excessive accumulation of cover-crop residue or soil should be removed. Fertilizer should be applied as needed to maintain a vigorous, dense growth of grass. If small washes occur in the waterway, they should be repaired by transplanting sod. The waterway should be inspected several times a year, preferably after a heavy rain, to determine maintenance needs.

UNDERGROUND OUTLETS

Underground outlets are used to dispose of runoff from terraces or from earth embankments used to stabilize natural depressions on unterraced land. Runoff is temporarily stored by the terrace or embankment and gradually drained off through an underground conduit. Because of the detention storage, only a small diameter conduit is needed.

DESCRIPTION

The underground outlet consists of vertical intake risers connected to the conduit (Fig. 6.17). Water flows into the outlet through holes

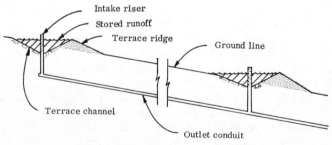

Figure 6.17. Underground outlet.

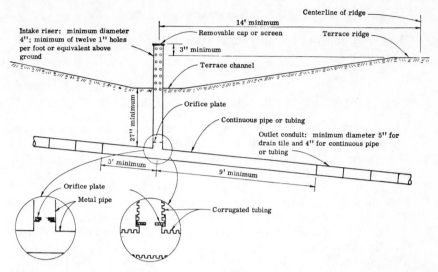

Figure 6.18. Details of the intake riser and the outlet conduit under a terrace. Reproduced from Otto Griessel and R. P. Beasley, Design criteria for underground terrace outlets, Univ. Mo. Ext. Div. Sci. Tech. Guide 1525, 1971.

in the riser. An orifice plate in the bottom controls the flow to the conduit so that there is no reverse flow in the intake risers at a lower elevation. Details of the intake riser and conduit are shown in Figure 6.18.

The intake riser may be connected directly to the outlet conduit, as shown in Figure 6.18, or it may be offset to one side and connected to the outlet by a short section of conduit. If offset, there is less danger of damaging the outlet conduit if the riser is hit by farm machinery.

The underground outlet is usually placed in a natural depression. The terrace channel or the land above the embankment is graded to the intake riser. The terrace or embankment top is constructed level and with sufficient height to provide storage for the maximum runoff expected in a ten-year period.

HISTORY

A sketch of a tile outlet terrace system was shown in the USDA Farmer's Bulletin 512 dated April 1917 (11). Very little channel storage was provided in these early systems and large conduits were required to carry the runoff. Only a limited number were installed because of the cost of the large conduit.

The present concept of using channel storage to reduce conduit size was developed by U.S. Soil Conservation Service engineers in Iowa in the early 1960s (10, 11).

ADVANTAGES

Underground outlets provide the following advantages in the disposal of runoff:

1. Better alignment of terraces is permitted because they may be built straight across depressions rather than curving up to cross them on grade.
2. The need for grassed waterways which take land out of production and obstruct farming operations is eliminated.
3. More cropland is provided where the landowner has no need for the grass produced in the waterways.
4. An outlet for underground drainage is provided where needed.
5. Peak rates of runoff from the area are reduced.
6. The amount of sediment in the runoff water is reduced.
7. The need for a stabilization structure at the lower end of a waterway may be eliminated.
8. The topography of the area may be improved. Soil which is eroded from the area will be deposited in the depressions, thus reducing the variation in land slope.

LIMITATIONS

There are however the following limitations in disposal of runoff by underground outlets:

1. Much more effort is required for design and layout than for grassed waterways.
2. Equipment that normally is not used in constructing a terrace system is required to install the conduit.
3. An appreciable amount of hand labor is necessary for installation.
4. The initial cost is more than for a grassed waterway unless a stabilization structure can be eliminated.
5. Crop growth may be restricted and field operations delayed on less permeable soils due to water stored in the terrace channel or above the embankment.
6. Holes in the riser or orifice may occasionally clog with debris and cause overtopping of the terrace or embankment.
7. The drainage area of the outlet system may be limited to about 20 acres; for larger drainage areas, the size of the conduit becomes larger and rather expensive.
8. Velocities in the conduit may become excessive on steep grades and cause movement of soil near the openings between field drain tile.

DESIGN

The combination storage capacity and outlet discharge is usually designed to handle the maximum amount of runoff expected from a six-hour storm in a ten-year period. A runoff of 2 inches from watersheds in row crops has commonly been used for design in locations along line 1.0 in Figure 5.9. If specific information on depth of runoff is not available for other areas, it may be determined by multiplying 2.0 inches by the location factor L (Fig. 5.9).

Intake Riser. The intake riser can be made of either smooth or corrugated pipe. It should be located in the lowest place in the channel to provide positive drainage, and should be a minimum of 14 feet from the center line of the embankment ridge. This distance will vary upward, depending on the width of the farm equipment that is to pass between it and the top of the ridge. The joint between the intake riser and the tee on the outlet conduit should be slightly flexible to prevent breaking the joint in case the riser is hit by farm equipment. The top of the riser should be at least 3 inches higher than the top of the ridge. It should be covered with a durable cap or screen which is removable so the riser can be cleaned in case the orifice plate becomes clogged. The openings in the screen should be no larger than the holes in the side of the riser.

Orifice Plate. A specifically designed orifice plate at the bottom of the riser controls the discharge. Table 6.7 gives the discharge rates of orifices having specific diameters acting under given heads and indicates how the head is to be determined. Orifices less than 1.5 inches in diameter are subject to clogging with debris and should not be used.

The orifice capacity should normally be not less than 0.042 cubic foot per second per acre of drainage area. This is equivalent to removing 2 inches of runoff in forty-eight hours. This rate of removal has proved adequate for most crops and soils where underground outlets have been used. If used in an area of less permeable soil and for a crop that is susceptible to flooding, a more rapid rate of removal may be desirable. For example, if it were considered desirable to remove 2 inches of stored water in twenty-four hours, the orifice capacity should be increased to 0.084 cubic foot per second per acre of drainage area.

If the storage in the terrace channel is less than that required to store the runoff expected from a six-hour storm in a ten-year period, the orifice capacity must be increased to prevent the embankment from being overtopped. For example, assume that an underground outlet is to be used in an area where 2 inches of runoff can be expected in a six-hour period, and storage is provided for only 1 inch of runoff. If the additional 1 inch is removed in the six-hour period, it will require an orifice capacity of 0.168 cubic foot per second per acre of drainage area. The entire 2 inches of runoff will then be removed in approximately twelve hours.

The orifice plate should fit tightly against the seat at the bottom of the riser. It should be made of durable material that will withstand the

TABLE 6.7. Discharge Rate for Circular Orifices in cfs

ORIFICE DIAMETER IN INCHES	HEAD IN FEET [1]						
	2.5	3.0	3.5	4.0	4.5	5.0	5.5
1.50	.094	.103	.111	.119	.126	.133	.139
1.75	.128	.140	.151	.162	.172	.181	.190
2.00	.166	.182	.197	.210	.223	.235	.247
2.25	.210	.231	.249	.266	.282	.298	.312
2.50	.259	.284	.307	.328	.348	.367	.385
2.75	.314	.344	.371	.397	.421	.444	.466
3.00	.373	.408	.441	.472	.500	.527	.553
3.25	.438	.479	.518	.554	.587	.619	.649
3.50 [2]	.508	.556	.600	.642	.681	.718	.753

[1] Head = $0.7h_1 + h_2$. See diagram below.
[2] A special design of the intake riser is required for orifice diameters over 3.50 inches.

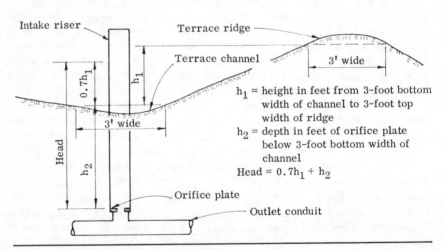

h_1 = height in feet from 3-foot bottom width of channel to 3-foot top width of ridge

h_2 = depth in feet of orifice plate below 3-foot bottom width of channel

Head = $0.7h_1 + h_2$

Source: Otto Griessel and R. P. Beasley. Design criteria for underground terrace outlets. Univ. Mo. Ext. Div. Sci. Tech. Guide 1525, 1971.

pressure and flow of water and should have a smooth finish so as not to catch debris.

Outlet Conduit. The outlet conduit may be made of clay or concrete drain tile, clay sewer pipe, plastic pipe, corrugated plastic tubing, smooth iron pipe, corrugated steel or aluminum pipe, or other material substantial enough to withstand loads applied by the earth cover and farm machinery. A section of this conduit under the impounded water should be of continuous pipe or tile with sealed joints. The upper end of the conduit should be sealed with a durable material.

The capacity required for any section of conduit is determined by adding up the discharge through the intake risers above that section. The size of conduit is determined by use of Figures 6.19 or 6.20. The velocity in the conduit should not exceed that specified in Table 6.8.

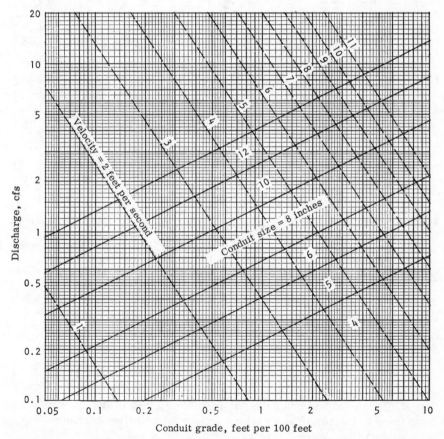

Figure 6.19. Discharge curves for outlet conduit. (Smooth pipe or tile having good alignment, n = 0.011.) (Courtesy U.S. Soil Conservation Service)

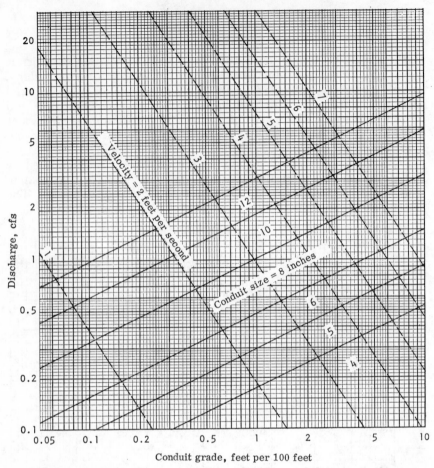

Figure 6.20. Discharge curves for outlet conduit. (Corrugated plastic tubing or tile not having good alignment, n = 0.016.) (Courtesy U.S. Soil Conservation Service)

Outlet Section. The outlet section of conduit should be metal pipe with an inside diameter equal to or greater than that of the conduit. Other details for the outlet section are given in Figure 6.21.

Embankment. The embankment to impound the water may be a section of a broad-base, steep-backslope, or flat-channel terrace. The top of the terrace is constructed level for that section designed to impound water. If the field is not terraced, an underground outlet may be used to dispose of the runoff water which accumulates in the depressions. The embankment constructed across the depression may have farmable front-slopes and backslopes or it may have a steep backslope.

TABLE 6.8. Maximum Permissible Velocities for Outlet Conduits in Feet per Second

Soil Texture	Drain Tile with Unsealed Joints	Drain Tile with Wrapped Joints[a] or Bell-End Tongue and Groove Pipe	Continuous Pipe or Sewer Pipe with Sealed Joints
Sand and sandy loam	3.5	6.0	No limit
Silt and silt loam	5.0	6.0	No limit
Silty clay loam	6.0	7.0	No limit
Clay and clay loam	7.0	9.0	No limit

Source: Otto Griessel and R. P. Beasley. Design criteria for underground terrace outlets. Univ. Mo. Ext. Div. Sci. Tech. Guide 1525, 1971.

[a] The tile should be laid with tight fitting joints and each joint entirely wrapped with a durable material.

Storage. The embankment should be built to the height necessary to provide the required storage. The height required is determined by trial and error. The height of the level section of the embankment ridge, which appears to be required to provide the storage needed, is selected. The storage is computed and if it is greater or less than the required storage, a new height of embankment is selected and the storage again computed. This operation is continued until a height of embankment is selected which provides the required storage.

In computing the volume of storage, channel cross sections are normally taken at 50-foot intervals and the average end area used to compute the storage volume. (See Ch. 13.) Various agencies have devel-

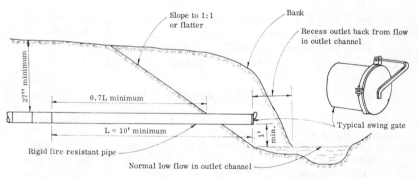

In order to protect the outlet end of the pipe from damage due to high flood flows in deeper channels, the outlet section should be modified as follows: (1) The minimum length of the outlet section should be at least equal to the depth of the channel, but not less than 12 feet, (2) the outlet section need not be recessed back into the bank, but a minimum of 4 feet of the pipe should extend beyond the bank, (3) the bottom of the outlet section should be a minimum of 1 foot above maximum flood stage in the channel, (4) no swinging gate is needed.

Figure 6.21. Details of the outlet section. Reproduced from Otto Griessel and R. P. Beasley, Design criteria for underground terrace outlets, Univ. Mo. Ext. Div. Sci. Tech. Guide 1525, 1971.

oped tables which can be used to make a rapid determination of storage volume.

Channel Grade. If the system is to be designed so that erosion does not occur in the terrace channel, the grade for that section outside the impounded area can be obtained from Chapter 7. The grade in the portion of the channel that temporarily impounds water can be increased up to 2 percent.

On deep, uniform soils a system may be designed so that soil eroded from the slopes will fill in the depressions at the intake, thereby improving the topography and farmability of the area above the terraces. In the design of this system there would be no limitation on the maximum grade of the terrace channel sloping toward the intake risers.

CONSTRUCTION

Care should be taken to install all parts of the underground outlet at the correct location and grade. Drain tile should be installed in good alignment and close fit. Earth backfill should be placed in such manner that the conduit is not displaced. Backfill material should be placed in 6-inch layers and hand tamped to a depth of 1 foot over the top of the conduit. The remainder should be backfilled and compacted by machine. That portion of the trench located under the storage channel and embankment should have the sides of the trench inclined to a slope of one to one or flatter before backfilling by machine.

MAINTENANCE

Maintenance will include removing debris from the holes in the intake risers and from the orifice plates, filling holes developed by water seeping from the storage channel to the conduit, and repairing rills in the embankment caused by overflow of excessive runoff.

PROBLEMS

6.1. A grassed waterway is to be constructed to carry 80 cfs. The slope of the waterway at the point being considered is 8 percent. It is on a silty clay soil, and an excellent stand of fescue will be maintained. Design the waterway—
    a. With a parabolic cross section. Compute the additional width required for rank vegetation using five-to-one side slopes.
    b. With a trapezoidal cross section with four-to-one side slopes.

6.2. A terrace channel is to be constructed to carry 10 cfs. The channel is to be cultivated and built on a sandy clay soil. It is to have a triangular cross section with ten-to-one side slopes.
    a. What is the maximum permissible velocity in cfs?
    b. What is the cross-sectional area of the channel in feet?
    c. What will be the depth of flow in feet?

d. What is the maximum slope that could be used without having excessive erosion when there is no vegetation in the channel?

e. If the terrace were built to the slope found in (d), what would be the maximum depth of flow if the runoff of 10 cfs occurred with grain sorghum drilled in 40-inch rows in the channel?

6.3. A parabolic-shaped waterway is to be constructed on a silty clay soil to serve as a terrace outlet. Terraces will discharge into the outlet from both sides. A good stand of fescue will be maintained in the outlet. A depth of cut of 0.8 foot will be made in constructing the terraces. The outlet will be crossed when farming the field. The slope in the upper end of the outlet is 10 percent, and the peak rate of runoff will be 10 cfs. The slope in the outlet gradually decreases to 6 percent at the lower end where the peak rate of runoff will be 40 cfs. The outlet will be 1,000 feet long. Give the design width and depth for each end of the outlet. Show by sketches similar to Figures 6.2 and 6.7 the size and shape of outlet you would recommend for construction, including all dimensions indicated.

6.4. A trapezoidal-shaped waterway is to be located along one side of a square 40-acre field to serve as an outlet for the terraces. A depth of cut of 1.0 foot will be made in constructing the terraces. The outlet will not be crossed in the farming operations. The soil is a silty loam, and a good stand of brome will be maintained in the outlet. The slope in the upper end of the outlet is 3 percent, and the peak rate of runoff will be 15 cfs. The slope gradually increases to 5 percent at the lower end where the peak rate of runoff will be 90 cfs. The land slope perpendicular to the outlet is fairly level throughout its length. Give the design width and depth for each end of the outlet. Show by sketches similar to Figures 6.1 and 6.7 the size and shape of outlet you would recommend for construction, including all dimensions indicated.

6.5. A trapezoidal-shaped waterway is to be constructed along a straight property line. It will carry the runoff from terraces and diversions across some relatively flat and productive river bottom land. The soil at this location is a silty clay. A fair stand of fescue will be maintained in the waterway. A depth of cut of 1.0 foot will be made in constructing the terraces and diversions which discharge runoff into the waterway. A diversion carrying 36 cfs enters the waterway at the upper end (point A) where the grade in the waterway is 7 percent. Point B is located 600 feet below A. Terraces enter the waterway between points A and B. The total runoff in the waterway at B is 52 cfs and the grade in the waterway is 5 percent. Point C is located 100 feet below B. A diversion enters the waterway at C. The total runoff in the waterway at C is 80 cfs and the grade in the waterway is 5 percent. No additional runoff enters the waterway below C. Point D is located 300 feet below C. The total runoff at D is 80 cfs and the grade in the waterway is 1 percent. Point E is located 1,000 feet below D. The total runoff and waterway grade is the same as at D.

a. Compute the design width and depth of the waterway at points A, B, C, D, and E.

b. Give the width and depth of waterway that you would recommend for construction at points A, B, C, D, and E.

c. Show by sketches similar to Figures 6.1 and 6.7 the size and shape of the waterway you would recommend for construction. Include all dimensions.

6.6. An underground outlet is to be used to dispose of runoff from a terraced field located near line 1.0 in Figure 5.9. The drainage area of each terrace is as follows: terrace 1—1 acre, terrace 2—1.5 acres, terrace 3—2 acres, terrace 4—2 acres, and terrace 5—2.5 acres.

a. Determine the size of orifice to be used in each riser if storage is provided for 2 inches of runoff, the runoff is to be removed within forty-eight hours, and the head on each orifice is 4.5 feet.

b. The conduit will be on a 4 percent slope between terraces 1 and 5 and on a 0.5 percent slope below terrace 5. What size conduit will be required? The soil is a silt loam. Will the velocity be excessive for drain tile with unsealed joints?

c. If storage can be provided for only 1 inch of runoff in terrace 5, what size orifice should be provided? What size conduit will be required below terrace 5?

REFERENCES

GRASSED WATERWAYS

1. Chow, Ven Te. *Open channel hydraulics*. New York: McGraw-Hill, 1959.
2. Cowan, Woody L. Estimating hydraulic roughness coefficients. *Agr. Eng.*, vol. 37, no. 7, July 1956.
3. Cox, M. B., and Palmer, V. J. Results of tests on vegetated waterways and methods of field application. Okla. Agr. Expt. Sta. Misc. Publ. MP–12, January 1948.
4. Larson, C. L., and Manbeck, P. M. Improved procedures in grassed waterway design. *Agr. Eng.*, vol. 41, no. 10, October 1960.
5. Ree, W. O. Hydraulic characteristics of vegetation for vegetated waterways. *Agr. Eng.*, vol. 30, no. 4, April 1949.
6. Ree, W. O., and Palmer, V. J. Flow of water in channels protected by vegetative linings. USDA Tech. Bull. 967, February 1949.
7. Smith, D. D. Bluegrass terrace outlet channel design. *Agr. Eng.*, vol. 27, no. 3, March 1946.
8. U.S. Soil Conservation Service, USDA. Handbook of channel design for soil and water conservation. SCS–TP–61, June 1954.

UNDERGROUND OUTLETS

9. Griessel, Otto, and Beasley, R. P. Design criteria for underground terrace outlets. Univ. Mo. Ext. Div. Sci. Tech. Guide 1525, 1971.
10. Jacobson, Paul. New developments in land-terrace systems. Trans. Am. Soc. Agr. Eng., vol. 9, no. 4, 1966.
11. Phillips, R. L. Tile outlet terraces—history and development. Trans. Am. Soc. Agr. Eng., vol. 12, no. 4, 1969.
12. Phillips, R. L., and Beauchamp, K. H. Parallel terraces. USDA Soil Conserv. Serv., 1966.

7 ~~ Terraces

A terrace, according to Webster, is a
raised level space supported on one or
more sides by a wall or bank of turf.

Many of the terraces constructed by earlier civilizations were of this type. They consisted of flat areas built up behind vertical stone walls. These terraces formed a giant stairway up the slope, the farming being done on the tread of the stairs. Such terraces are called "bench terraces." They are still used where intensive cultivation of very steep land is required and provide farmland that could not otherwise be used because of the severe erosion hazard.

In the latter part of the nineteenth century, the farmers and conservationists in the southeast section of the United States began to realize that erosion was a serious problem. Many farmers constructed hillside ditches at intervals across the slope to remove the runoff water before it accumulated in sufficient volume to cause serious erosion. These ditches reduced erosion losses, but they resulted in an obstruction in the field and were difficult to maintain. In some areas the ditches were replaced by narrow levees, called narrow-base terraces, constructed across the slope. These terraces were too narrow and high to cross conveniently, so they too interfered with the farming of the fields.

In 1886 a farmer, Priestly M. Mangum, who lived near Wake Forest, North Carolina, plowed earth to the narrow levees and produced a broad ridge with a channel above. The side slopes on the channel and ridge were flat enough so that crops could be planted on and over them. This terrace became known as the Mangum, or broad-base, terrace and has been used extensively in this country. Over the years a number of different types of terraces have been devised to meet specific needs (5).

CLASSIFICATION OF TERRACES

Terraces may be classified by the method of disposal of the runoff, by the shape of the terrace cross section, or by alignment between terraces.

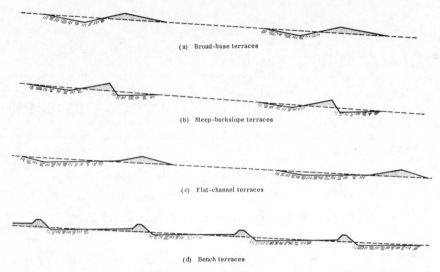

(a) Broad-base terraces

(b) Steep-backslope terraces

(c) Flat-channel terraces

(d) Bench terraces

Figure 7.1.   Types of terraces.

METHOD OF RUNOFF DISPOSAL

The graded terrace is constructed with the channel having a slight grade so that the runoff from the area above moves along the channel to an outlet. The graded terrace is used in areas where the rainfall and soil conditions are such that the runoff exceeds the moisture storage capacity of the soil. The primary function of the graded terrace is for erosion control; however, it does provide the opportunity for additional moisture conservation.

The level terrace is constructed with the channel level and the ends blocked so that runoff is stored in the terrace until it can infiltrate into the soil. The use of level terraces is limited to those areas of moderate

Figure 7.2.   Broad-base terrace. (Courtesy U.S. Soil Conservation Service)

to low rainfall and permeable soil. The runoff must be absorbed rapidly enough that crops in the terrace channel will not be damaged. The level terrace is used for erosion control and moisture conservation. Because these terraces hold surface runoff back until it is absorbed by the soil, they effectively reduce peak rates of runoff and flood damage.

In areas where the soil will not absorb water fast enough to prevent damage to crops in the channel, an outlet is provided for level terraces to increase the rate of water removal. Tile outlets may be used or the terraces may be constructed with open ends and vegetated outlets provided at these points.

SHAPE OF THE TERRACE CROSS SECTION

The broad-base terrace is constructed so that crops can be planted and farm machinery operated over the entire terrace (Figs. 7.1[a] and 7.2). It may be used either as a graded or level terrace. The earth to build the ridge is usually taken from the uphill side. This results in an increase in the slope of the land between terraces, since the difference in elevation from the top of the terrace ridge to the bottom of the terrace channel below is greater than the difference in elevation in the land surface at these points before the terraces were built. This type of terrace is best adapted to land slopes of less than 12 percent. As land slopes exceed 12 percent, it becomes increasingly difficult to farm on the backslope of the terrace ridge.

The steep-backslope terrace is constructed with a very steep backslope, two to one or steeper, which is kept in grass (Figs. 7.1[b], 7.3, and 7.4). It may be used either as a graded or level terrace. The earth to build the ridge may be taken from the lower side or from both the lower and upper sides of the ridge. Taking earth from the lower side of the ridge will reduce the slope of the land between terraces. However, a greater amount of earth will be required in the ridge to provide the needed capacity. Earth may also be taken from field ridges between terraces and used to make fills across depressions. This tends to make the

Figure 7.3. Steep-back-slope terrace. (Courtesy U.S. Soil Conservation Service)

Figure 7.4. *Steep-backslope terrace. (Courtesy U.S. Soil Conservation Service)*

Figure 7.5. *Parallel flat-channel terraces in southwestern Nebraska. The channels are 54 feet wide. Land slope is 2 percent. The photograph was taken after a 4-inch rain which fell during a one-hour period. (Courtesy U.S. Soil Conservation Service)*

land between terraces more farmable. As erosion occurs between terraces, the slope between will be reduced, the height of the terrace and the length of the backslope increased, and the width of the bench decreased. On slopes over 10 or 12 percent it is sometimes recommended that soil be plowed downhill between terraces so that the land slope will be reduced and the farmability improved in a shorter period of time (3, 4, 9). This terrace is adapted to steeper land on which it would be difficult to farm the backslope of a broad-base terrace. The backslope must be maintained in grass; and trees, weeds, and burrowing animals must be controlled. The area in grass on the backslope reduces the area of the field available for row crop production.

The *flat-channel terrace* is constructed with a wide flat channel which is made level and blocked at each end (Figs. 7.1[c] and 7.5). The width of the bottom of the channel may vary from 25 to 75 feet, depending on the slope of the land and the width of the implements used in farming the terrace. These terraces provide a large area for storage of surface runoff and are effective in retaining moisture for crop production and in reducing erosion and flooding. They are used primarily in areas of less than 25 inches of annual rainfall and are particularly effective where intense rains can be expected during the growing season. It is common practice to plant a crop in the channel which will benefit most from the additional moisture and to plant a more drought-resistant crop on the sloping areas between channels (2). This type of terrace is also known as the "Zingg conservation bench terrace."

The *bench terrace* is a modification of the earlier bench terraces (Figs. 7.1[d] and 7.6). It is constructed with a wide level bench separated by a steep slope which is kept in grass. They are used primarily for the irrigation of sloping land and are constructed with a slight grade along the bench to facilitate irrigation. They are also quite effective in the conservation of water and in the reduction of erosion from intense summer rainstorms. This practice is also called bench leveling and contour benching.

ALIGNMENT BETWEEN TERRACES

Nonparallel terraces are laid out without attempting to make them parallel. Some variation in grade or in the depth of cut may be made to reduce sharp curves along the terrace. They are best adapted to uniformly sloping land and to land which is to be used for crops such as small grain that do not require extensive machinery operations.

Parallel terraces are laid out with adjacent terraces spaced at an equal distance throughout their length, when possible. On irregular topography it may be necessary to divide the terraces into parallel sections with different spacings between. The grade in the terrace channel as well as the depth of cut may be varied to make the terraces or sections of terraces parallel (1, 9). Parallel terraces are more difficult to lay out

Figure 7.6. Bench terraces, or contour benches, in north-central Kansas, show-ing brome grass on the steep slope between benches. (Courtesy U.S. Soil Con-servation Service)

and usually cost more to construct. They are adaptable where the land is to be used for crops that require numerous machine operations in their production.

SOIL AND WATER CONSERVATION

Terracing is one of the basic practices in a successful plan for con-servation of soil and water on sloping land. It is not necessarily the com-plete solution to the problem but provides a basis or foundation for the effective use of agronomic practices. Terraces contribute to an integrated conservation plan in one or more of the following ways:

Division of the field into separate drainage areas. Each terrace re-ceives runoff only from the area between it and the terrace above, except for the top terrace which receives runoff from the area between it and the watershed boundary. Therefore there are as many independent drain-age areas as there are terraces across the slope. Since the runoff from each independent drainage area is intercepted by the terrace, excessive concentration of water does not occur on the lower portion of the slope.

Reduction of the length of slope. The most severe rill and gully ero-sion occurs on the lower portion of a slope. The runoff on the upper por-tion moves off predominantly as sheet flow. Water moving in sheet flow has a low velocity and limited eroding power. As the slope increases in length, there is an accumulation of runoff, which tends to concentrate in rills that grow into larger depressions until the sheet flow has changed to channel flow. Water moving as channel flow has a greater velocity with greatly increased eroding power. The distance between terraces should be short enough so that the runoff will move to the terrace channel primarily as sheet flow.

Reduction of the need for drainage and improvement of the productivity of bottomland. In many cases water from hill land discharges onto flatland below, causing overflow and deposition problems, and increases the need for drainage on these fields. If drainage ditches are built to remove water from the bottomland, they must be made large enough to carry this excess runoff, and in many cases they will be clogged with silt washed in from the eroding hillside. Terraces on the adjacent rolling land will divert the water from the bottomland, thus improving its productivity and at the same time reducing the drainage problems.

Solution of combined drainage and erosion control problems. A combination of erosion and drainage problems exists in many areas where the topsoil is underlain with a relatively impermeable subsoil and where the slopes are long and relatively flat. These soils dry out slowly, delaying seedbed preparation and planting, resulting in mired machinery, reduced yields, and often delayed harvest. The runoff rate from this soil of low permeability will be high; and even though the slopes are relatively flat, serious erosion will result. Terraces or cross-slope channels can be used to provide drainage and to prevent the accumulation of runoff on the lower part of the slope.

Contribution to water conservation. Level terraces are particularly effective in the conservation of water and in the reduction of surface runoff, especially so in areas of limited rainfall and where rain comes in a few intense storms during the growing season. Crop rows planted across the slope with terraces will also impound runoff and provide additional time for water to infiltrate into the soil. The delay of runoff water in the channel of the graded terrace also gives time for increased infiltration into the soil. However on land with an impervious subsoil the amount of infiltration in the channel will be slight and the conservation of water limited. On deep, open soil an appreciable saving is possible. This retention of the water on the land will provide more moisture for crop growth which will result in larger yields and higher income. There will also be less surface runoff and a reduction in flood damage.

Reduction of flood runoff and damage. Level terraces provide storage for practically all the surface runoff from ordinary rains and effectively reduce the surface runoff from extreme rains. Graded terraces delay the time required for the runoff to leave the land and increase the amount of infiltration, thereby decreasing the peak rate of flow, particularly on small watersheds. Terraces further reduce flood damage by holding the soil on the land so that it will not be eroded downslope to fill channels and reservoirs.

Greater liberty in selecting farm enterprises. Crops vary in the amount of net income they provide for the farm business. Unfortunately, the crops producing the highest income per acre often provide little protection against erosion. For example, corn and soybeans are high-income-producing crops in the Corn Belt, as are cotton and tobacco in the southern states. If a farmer attempts to grow these crops on rolling land, serious soil depletion results, and he finds it difficult to economically maintain production. If terraces are used to control erosion, these high-income-producing crops can be grown, and the productivity of the soil is maintained. If terraces are not used, the farmer must limit the crops grown to those that give more protection from erosion and must accept a reduction in income. Thus terraces not only give protection from erosion but also allow the farmer more flexibility in selecting the farm enterprise.

Reclamation of eroded land. It has been estimated that there are 50 million acres of land in the United States that have been stripped of topsoil or so riddled with gullies as to make them virtually useless in their present state. If this land is to be placed in production, the erosion must first be controlled. The first step in the process is to fill the gullies, which destroys the drainage pattern that has developed. The second step is to terrace the land, thus controlling erosion and preventing the reestablishment of the previous drainage pattern. The cost is high, but if this land is needed for crop production, erosion must be controlled.

Preservation of better land. Often a farmer will not consider terracing a field until it is eroded to the point that gullies seriously interfere with the operation of farm machinery and production drops because of the loss of topsoil. It will be difficult if not impossible to construct a good system of terraces on this badly eroded land. The terraces will be crooked, unevenly spaced, and difficult to farm. Also the cost of construction will be much greater on the eroded land. The terraces in themselves will not improve the productivity of this seriously eroded land; this must be accomplished with good soil and crop management practices. Large expenditures for fertilizer will be needed to replace the nutrients lost by erosion. It is imperative that erosion be controlled before it is allowed to progress to this stage.

There are extensive areas of land that are subject to erosion but still have a reasonable amount of topsoil on which serious gullying has not developed. This better land should be terraced while it is still possible to lay out a good system at a reasonable cost and while the field is still productive. With erosion controlled, this area can be cropped intensively to high-value crops. The more severely eroded areas should then be used for less erosive crops until the demand for additional cropland necessitates their reclamation.

Continuous protection. Most crops furnish protection from erosion when they are fully developed; but there are periods during seedbed preparation, during early stages of growth, and after harvest when the protection they offer is inadequate. Terraces furnish the needed protection during these periods as well as during all other periods of the year.

Protection of investments in soil improvement. Large expenditures in time, labor, fertilizer, and lime are often involved in maintaining high production. Should an intense rain occur when the soil is not protected, a serious loss of soil and of these investments results. Terracing is a way of insuring these investments.

Terraces thus contribute in many ways to the conservation of soil and water resources. It will never be possible to eliminate soil erosion completely, but it is possible by the use of an integrated program to reduce erosion losses to a point where productivity can be maintained or increased.

OBJECTIVES

The following objectives should be kept in mind in planning a terrace system: erosion control, moisture conservation, farmability, and improvement of topography (9). Many of the older terrace systems were planned with erosion control as the prinicipal and often the only objective, and many times they were difficult to farm and maintain. A well-planned system must not only control erosion and conserve moisture but must result in an improvement of the topography and in the farmability of the field.

EROSION CONTROL

The terrace system should be designed to control erosion under the most intensive use to be expected on the land. It would be a mistake to base the design on the present use if in a few years much more intensive cropping is to be expected.

MOISTURE CONSERVATION

Moisture deficiencies and unfavorable distribution of precipitation are major problems in crop production in many areas. Here, terraces should be designed to store and make maximum use of the rainfall which does occur.

FARMABILITY

The following quotation expresses the need for designing a terrace system that is farmable: "Of what value is the best system of erosion control in the world if the majority do not accept it? Terraces are one of the best erosion control practices we have, yet they are not accepted by many.

To be accepted, today's terrace systems must be as modern as the equipment which farms the land, as valuable as the soil they protect, and as desirable as we have the know-how to make them. And above all, they must be farmable" (9).

Terraces which are farmable—

1. Are parallel wherever possible. This eliminates point rows and turning between terraces. It saves time in plowing, planting, and harvesting and reduces crop damage and soil compaction resulting from turning between terraces.
2. Have the curvature of the terraces reduced to the point that equipment can easily farm the crop rows parallel to the terraces.
3. Are spaced to fit the equipment which will be used in farming them in the future. If possible, a spacing should be selected which will fit various row widths and multiples of row widths.
4. Are constructed with a cross section that is easy to farm. The side slopes of the terrace channel and ridge should be flat enough and wide enough to permit ease of operation of the farm equipment.
5. Provide convenient access to all fields to facilitate the movement of livestock and machinery from the farmstead to the fields and from field to field with a minimum of travel and a minimum of gates to open.

IMPROVEMENT OF TOPOGRAPHY

Terrace systems should be so designed that the movement of soil in the farming operations and by erosion that takes place after the terraces are constructed will result in a topography that is easier to farm and less subject to erosion.

Many fields are badly eroded before they are terraced. The curves in such terraces should be straightened as much as possible by cutting and filling during construction. The irregular land surface between terraces can be improved by subsequent farming operations. In some cases it is advisable to use a land plane to smooth the land between terraces. The land smoothing will reduce the concentration of water between terraces, thereby reducing erosion.

On extremely irregular topography and very deep, erodible soils, Jacobson (4) recommends a method of progressive terracing from the bottom of the hill to the top. A terrace is constructed at the bottom of the slope and deep fills are made across the depressions with soil taken from the ridges, so that the terrrace is relatively straight. Underground tile outlets are provided in the depressions to remove the runoff. Soil that is eroded from the hillside above the terrace will be deposited in the depressions and in this way the topography above the lower terrace will be smoothed out. When the topography above the bottom terrace is improved sufficiently to be quite farmable, the next terrace up the hill is constructed and the process repeated.

DESIGN

When a farm is terraced, all farming operations in the future must be accomplished with the terraces on the land. Since terraces must be considered as a permanent improvement, it is extremely important that they be designed and laid out in such manner to provide the best possible soil and water conservation and offer the least hindrance to the farming operations.

Terrace design involves the proper spacing of the terraces on the slope, the selection of a channel with proper grade and adequate capacity for runoff, and a cross section that can easily be farmed with modern machinery.

SPACING

Broad-Base Graded Terrace. It is desirable to space terraces as far apart as possible. Wide spacing makes it easier to operate machinery in the field and reduces the cost of terracing. However, a number of factors limit the spacing between terraces: (1) erosion increases with length of slope; (2) if water concentrates between the terraces, small gullies may form and cause deposition in the channel of the terrace below; and (3) runoff from the wide spacing may overload the channel of the terrace, causing overflow and damage to the terrace involved and to the land below.

The principal factors affecting terrace spacing are the land slope, soil type, intensity of rainfall to be expected, crops to be grown, and machinery that will be used in farming. The terrace spacing for different sections of the country has been determined by research and field experience. Equation (7.1) may be used as a general guide for terrace spacing if more specific guides are not available (10).

$$VI = XS + Y \qquad (7.1)$$

where $VI =$ vertical interval between terraces, in feet (Fig. 7.7).

$X =$ a variable depending upon geographic location (Fig. 7.8).

$S =$ the average slope of the land draining into the terrace in feet per 100 feet, percent slope.

$Y =$ a variable with values ranging from 1.0 to 4.0 depending on soil erodibility, cropping systems, and crop management systems. The lower value is applicable for very erodible soils with conventional tillage methods where little residue is left on the surface. The higher value is applicable to erosion-resistant soils where tillage methods are used that leave a large amount of residue (minimum of 1.5 tons straw equivalent) on the surface.

The vertical interval, or difference in elevation between terraces, is used for the layout of nonparallel terraces because on irregular topography the surface distance between terraces varies depending upon the land slope.

In the layout of parallel terraces the horizontal interval (Fig. 7.7) is used in spacing terraces or sections of terraces. The recommended horizontal interval may be computed by Equation (7.2) after the vertical interval for the terrace has been determined.

$$HI = \frac{VI}{S} \times 100 \tag{7.2}$$

(Parallel terraces are usually spaced by measuring the surface distance between them rather than the horizontal distance. For land slopes on which most terraces are constructed there is no significant difference between the horizontal and surface distance. For example, on a 14 percent land slope this difference is only 1 percent.) The vertical interval that will result in the correct horizontal interval between terraces can be determined by Equation (7.3).

$$VI = \frac{HI \times S}{100} \tag{7.3}$$

In the design of terraces the size of equipment now being used as well as the equipment that may be used in the future should be considered. For example, four-row equipment with 40-inch row spacing is quite common today but the trend in the future may be to 30-inch rows with six-row or eight-row equipment. If possible a spacing should be selected which will be adaptable to a wide variety of row and machinery widths.

The spacing given by Equations (7.1) and (7.2) may be varied up to 20 percent to miss obstacles in the field, to reach a satisfactory outlet, or if terraces or sections of terraces are to be made parallel, to accom-

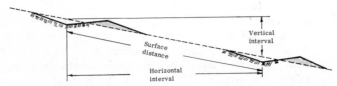

Figure 7.7. Measurements used in spacing terraces.

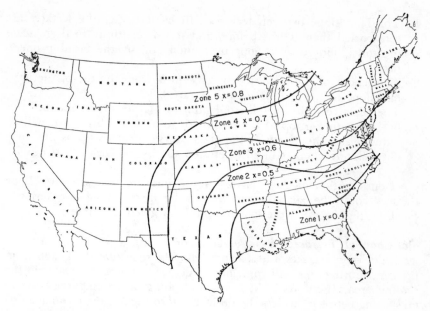

Figure 7.8. Values of X *in the equation* VI = XS + Y.

modate the width of machinery that will most likely be used in farming the terraces.

On steeper slopes, the use of Equation (7.1) will give a terrace spacing which is too narrow for efficient operation of large equipment. Most states recommend that a certain minimum spacing be used no matter how steep the land slope. In many of the Corn Belt states this minimum is 80 feet. In order to keep erosion within acceptable limits, it may be necessary to include some grass in the rotation on these wider spacings on the steeper slopes.

The spacing between terraces should not exceed the slope length determined for contour cultivation by using the soil-loss prediction equation (see Ch. 4).

Broad-Base Level Terrace. In areas of limited rainfall where conservation of moisture is the primary objective, the spacing of level terraces is a function of the infiltration rate in the channel and of the runoff. In more humid areas where erosion control is a principal objective, the level terraces should be spaced approximately the same as graded terraces.

Steep-Backslope Terrace. Steep-backslope terracing is a relatively new practice in the United States, and general specifications for spacing have not been developed. Some factors to be considered in spacing are—

1. The land slope between terraces will be less after the terraces are constructed than it was before, and it will continue to decrease as the soil is moved downslope by erosion and by the farming operations.
2. The length of slope affected by erosion is from the toe of the grass backslope to the channel of the terrace below.
3. The effective width of the cropland area is from the toe of the grass backslope to the top of the terrace ridge below. This width will be reduced and the length of the grass backslope increased as the slope between terraces is reduced by erosion and by the farming operations.

Consideration of these factors would indicate that steep-backslope terraces could be spaced somewhat wider than broad-base terraces. Jacobson (6) suggested the spacings for Iowa given in Table 7.1.

Flat-Channel Terrace. If conservation of moisture is the primary objective of these terraces, the spacing should be such that the runoff from the contributing area will contribute a desirable amount of water to the channel area. If erosion is the factor controlling the spacing, the horizontal spacing between terraces should be equal to the spacing recommended for broad-base terraces plus the width of the channel. If the flat-channel terraces are to be made parallel, the horizontal interval between them can be determined by

$$HI = \left(\frac{VI}{S} \times 100 \right) + \text{width of channel} \qquad (7.4)$$

If the flat-channel terraces are not to be made parallel, the vertical interval will be more convenient to use in layout. This can be computed by Equation (7.3).

TABLE 7.1. Spacing of Steep-Backslope Terraces

Slope (%)	Spacing (ft)	Bench Width (ft)	
		Initial	Final
2	245	240	240
4	245	240	240
6	171	160	160
8	137	120	120
10	142	120	120
12	102	80	80
14	108	80	80
16	113	93	80
18	118	107	80
20	126	107	80

Source: Paul Jacobson. Improvements in bench terraces. Trans. Am. Soc. Agr. Eng., vol. 11, no. 4, 1968.

CHANNEL GRADIENT

The gradient in the terrace channel must be sufficient to prevent ponding but not so great as to cause erosion in the channel.

Minimum Grade. It is difficult to construct the terrace to a grade more accurate than plus or minus 0.10 percent, and tillage operations will tend to build up obstructions in the channels. Sufficient grade must be given so that ponding in the channel will not seriously damage crops or delay field operations. Suggested minimum grades are 0.2 percent for soils of low permeability, and 0.0 percent for soils of high permeability.

Level terraces are constructed with zero grade in the channel. However it is sometimes desirable, in order to make terraces parallel, to split the terrace into sections and use a different channel elevation in each. The ends of the sections are blocked with earth to prevent water from flowing to the lower section.

Maximum Grade. If the terrace is constructed with too much grade, the increased velocity will erode the channel, carry soil from the field, and cause destructive deposition in the terrace outlet. The maximum grade that can be used safely will vary with the erodibility of the soil and the depth of flow in the channel. Recommendation R268 in the ASAE Yearbook (10) specifies that the maximum velocities should be 2.5 feet per second for soils highly resistant to erosion, 2.0 feet per second for most soils, and 1.5 feet per second for extremely erodible soils. Velocities are computed by Manning's formula using an n value of 0.03. Table 7.2 may be used to obtain maximum grades in different sections of a typical broad-base terrace for most soils if velocities are not computed by Manning's formula.

TABLE 7.2. Maximum Grades for Different Sections of a Broad-Base Terrace

Distance from Upper End of Terrace[a] (ft)	Maximum Grade (%)[b]	Maximum Drop in a 50-Foot Interval (ft)
0–50	2.4	1.20
50–100	2.0	1.00
100–150	1.6	0.80
150–200	1.2	0.60
200–250	1.0	0.50
250–300	0.8	0.40
300–350	0.7	0.35
350–450	0.6	0.30
450–550	0.5	0.25
550–1,200	0.4	0.20
1,200–1,600	0.3	0.15

[a] Assuming a drainage area equivalent to a normal terrace spacing for this length.
[b] If the terrace discharges into a grassed outlet, the grade in the 50-foot section next to the outlet should not exceed 0.4 percent or 0.3 percent if the length of the terrace exceeds 1,200 feet.

Selecting the Grade. Several variations in grade may be used in designing a terrace.

Variable grade. The grade may be varied in any section of a variable-grade terrace between the minimum and maximum grade. The advantage of using the variable grade is that the terraces can be made more nearly parallel at a minimum cost.

Increasing grade. The grade is varied from a minimum at the upper end to a maximum at the outlet end of the terrace where the greatest amount of runoff occurs. The minimum grade at the upper end provides for the maximum absorption of runoff but may result in ponding problems in wet seasons. The increasing grade is not adaptable to parallel terrace layout.

Constant grade. The grade is held constant throughout the length of the constant-grade terrace. It is the easiest type to lay out and construct, but does not provide for the increased absorption of moisture at the upper end, nor is it adapted to parallel terrace layout.

CROSS SECTION

A terrace cross section should have adequate capacity to carry the runoff; it should be designed to fit the topography, the farm machinery to be used, and the crops to be grown; and it should be economical to construct with the equipment available.

Broad-Base Graded Terrace. A typical broad-base terrace cross section is shown in Figure 7.9. The terrace may be constructed with a V-shaped cross section or with a flat bottom. The V-shaped channel is more desirable on less permeable soils where channel wetness would be a problem in the flat bottom. The flat bottom may be used on more permeable soils where wetness is not a problem. The size and type of machinery to be used in farming the terraces should also be considered in selecting the terrace cross section.

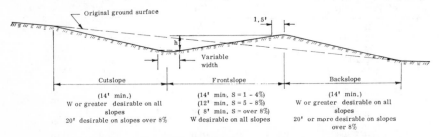

Figure 7.9. A typical broad-base terrace cross section.

The amount of earth movement required to build a broad-base terrace increases with the land slope. For example, a V-shaped terrace with a 14-foot cutslope, frontslope, and backslope, and a 1.5-foot height requires the following amount of earth to be moved on different land slopes: 13.8 cubic feet per foot on a 4 percent slope, 17.6 cubic feet per foot on an 8 percent slope, and 21.5 cubic feet per foot on a 12 percent slope.

The construction of broad-base terraces results in an increase in the land slope between terraces. After terraces are constructed, the new land slope is measured from the top of the terrace ridge to the channel of the next terrace downslope. If terraces with a 1.5-foot ridge height and 14-foot cutslope, frontslope, and backslope are constructed on a 12 percent land slope at a spacing of 80 feet between terraces, the effective land slope between terraces would be increased from 12 percent to 17 percent. The backslope of these terraces would be 24 percent, the frontslope 11 percent, and the cutslope 24 percent.

Because of the increased earth movement and the steep cutslope and backslope which make farming quite difficult, terraces on slopes greater than 8 to 10 percent are often constructed with a shorter frontslope. If the frontslope were 8 feet and the cutslope and backslope 14 feet, then the earth movement required to build a terrace with a 1.5-foot height on a 12 percent slope would be 14.8 cubic feet per foot of terrace. The land slope between terraces would be 15 percent. The backslope would be 21 percent, the frontslope 19 percent, and the cutslope 21 percent. This terrace with the narrower frontslope would cost less to construct and would have a decrease in the slope of the cutslope, the backslope, and the field slope. However, narrower equipment would be needed to farm on the 8-foot frontslope.

Height. Graded terraces are usually designed to carry the peak rate of runoff to be expected in a ten-year period. The depth of flow can be computed by Manning's formula, using a value for n of 0.06. A freeboard of 0.3 to 0.5 foot is added to the depth of flow to obtain the terrace ridge height, which is measured as indicated by h in Figure 7.9. The freeboard allows for irregularities in construction and reduction of the terrace height due to farming operations.

Table 7.3 gives the depth of flow and the ridge height for a V-shaped terrace with ten-to-one side slopes, a grade of 0.3 percent, and 0.5-foot freeboard, located near line 1.0 (Fig. 5.9).

Length. A terrace length of 1,600 feet is usually considered to be the maximum for graded terraces. If terraces of greater length are required, that section in excess of 1,600 feet should be designed as a diversion.

Broad-Base Level Terrace. The design of the level terrace differs from the design of the graded terrace in that the amount of runoff to be stored

TABLE 7.3. Depth of Flow and Terrace Heights

Distance from Upper End of Terrace (ft)	Depth of Flow (ft)	Height (ft)
100	0.3	0.8
200	0.4	0.9
400	0.5	1.0
600	0.6	1.1
800	0.7	1.2
1,000	0.8	1.3
1,300	0.9	1.4
1,600	1.0	1.5

determines the size. The terraces must be large enough to contain the design runoff volume without overtopping the ridge. In most areas the depth of runoff to be expected from a ten-year frequency, six-hour storm is used in determining the required storage capacity. The volume of runoff per foot of terrace will be equal to the terrace spacing multiplied by the depth of runoff. If specific information on depth of runoff is not available, it may be determined by multiplying 2.0 inches by the location factor L given in Figure 5.9. It is advisable to provide a freeboard of at least 0.3 foot above the depth required to contain the design runoff.

End blocks on level terraces should be designed so water will flow over them before overtopping the terrace ridge. Channel blocks at least 0.7 foot high should be spaced as needed to distribute the water throughout the channel.

Steep-Backslope Terrace. The frontslope of the steep-backslope terrace is made of sufficient width to fit the equipment to be used on the terrace. This width is seldom less than 14 feet. The backslope is usually constructed on a two-to-one slope or steeper. The earth to build the ridge may be taken from the lower side or both the upper and lower sides of the ridge. When all the earth is taken from the lower side, the ridge must be higher and more earth movement will be required (Fig. 7.10). When earth is taken from both sides of the ridge, some capacity is provided by the channel and less earth movement is required, but the land slope between terraces is increased (Fig. 7.11).

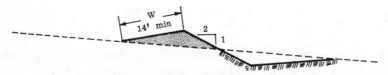

Figure 7.10. Cross section of a steep-backslope terrace with all earth taken from the downhill side.

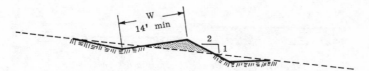

Figure 7.11. Cross section of a steep-backslope terrace with earth taken from both sides of the terrace.

Flat-Channel Terrace. The width of equipment to be used in farming the terrace and the land slope determines the dimensions of the terrace cross section. The ridge slopes should be wide enough to accommodate the widest equipment that is to be used. This width should seldom be less than 14 feet. The width of the channel is determined by the width of the equipment to be used and the land slope. The following widths are commonly used for the indicated slope range: with a land slope of 0 to 1 percent, the width of the channel is 60 to 75 feet; land slope 1.1 to 2.0 percent—channel 48 to 60 feet; land slope 2.1 to 3.0 percent—channel 36 to 45 feet; and land slope 3.1 to 4.0 percent—channel 24 to 30 feet.

The ridge should be high enough to contain the runoff to be expected from a ten-year frequency, six-hour storm. A ridge height of 1.3 feet is commonly used (Fig. 7.12).

Bench Terraces. The difference in elevation between the benches *d* in Figure 7.13 and the width of the bench *W* depends on the depth of soil, the width of equipment to be used in farming the benches, the type of equipment available to construct the benches, the land slope, and the amount the landowners are willing to spend.

Bench terraces are constructed primarily to provide for efficient irrigation of sloping upland. Additional information on design and layout can be found in literature on irrigation.

LAYOUT

Terraces should not be laid out until a water management plan for the entire farm or unit has been developed and the outlet location and type of outlet have been selected (Ch. 11).

NONPARALLEL TERRACES

No set procedure exists that will give the best terrace layout on all fields; consequently the final location of the terraces will depend to a

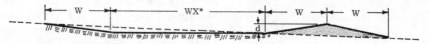

X* = Number of trips with equipment having a width W

Figure 7.12. Flat-channel terrace cross section.

Figure 7.13. Bench terrace cross section.

great extent on the judgment of the person doing the layout. The
following is a suggested procedure for those with little or no experience
in this area. As experience is gained and judgment is developed, modifications in this procedure to reduce the time required and to improve
the layout will be made.

1. Locate the high point of the area to be terraced. The level should
   be used to locate this point, for it is often impossible to judge the
   highest elevation in an area by the eye.
2. Determine the land slope above the first terrace. Measure this slope
   from the high point. In measuring land slope it is convenient to
   use a distance of 100 feet between points; however on steep land it
   may be advisable to use a 50-foot distance.
3. Select the vertical interval for the first terrace. Use the steepest
   slope from the high point and substitute in the equation $VI = XS + Y$.
4. Locate the first stake on the top terrace. This stake is usually located directly down from the high point. The difference in elevation between this point and the high point is equal to the vertical
   interval computed in (3).
5. Stake out the terrace to grade. Stakes are usually set at 100-foot
   intervals on uniformly sloping land with less than 5 percent slope. A
   50-foot distance is used on land with irregular slopes and land with
   slopes greater than 5 percent. If level terraces are being staked out,
   all stakes are set at the same elevation as the first stake. If graded
   terraces are being staked out, care should be taken to see that the
   rod reading is increased when approaching the outlet end of the
   terrace and decreased when approaching the upper end.
6. Check the location of the top terrace. Due to variations in topography on some fields it may be found that the top terrace, located as
   suggested above, will not be as near the recommended spacing in all
   parts of the field as would be desired. For this reason it is important
   to check the location of the top terrace before proceeding with the
   layout of additional terraces. To accomplish this, measure the land
   slope from the top of the ridge to the first terrace at additional
   points in the field. Compute the recommended vertical interval or
   surface distance for terraces on these slopes. Measure either the
   vertical interval or the surface distance between the top of the ridge
   and the top terrace at the points where the land slope was measured
   and compare to the recommended interval or distance.

7. Shift the location of the top terrace, if necessary, to give a more desirable spacing. In some cases the top terrace may be moved downslope to give a more desirable spacing in some sections of the field even though this may result in a spacing that is wider than recommended in another section of the field. In general, a somewhat wider spacing above a certain section of a terrace is permissible if the area involved is limited, the topography of the area is such that the runoff water will be spread over a wider area rather than being concentrated, and the wide spacing is near the upper end of the terrace.

8. Measure the slope of the land draining into the second terrace. A sufficient number of slope readings should be taken to determine the average slope of the land draining into the second terrace. If the terrace is short and the slope uniform, one reading would be adequate. If however the terrace is long and the slope irregular, several readings will be necessary to obtain an average.

9. Select the vertical interval for the average slope. Use the average slope and substitute in the equation $VI = XS + Y$.

10. Locate the first stake on the second terrace at the given vertical interval below the first terrace. If the second terrace will be the same length as the first terrace and is to be staked to the same grade, the vertical interval between the two terraces will be the same at all points along them, in which case the first stake on the second terrace can be located below any point on the first terrace. If the second terrace is not the same length as the first one, then judgment must be exercised in selecting the first point on the second terrace so that the most desirable spacing between them is obtained.

11. Stake out the terrace to grade.

12. Additional terraces are located in the same manner as the second terrace.

The spacing of terraces is based on the average land slope between them. For example, if land slopes between terraces were 4, 6, and 8 percent in different parts of a field, the terrace spacing would be based on the average slope of 6 percent. The vertical interval for a 6 percent slope, using the equation $VI = 0.6S + 2$, would be 5.6 feet. The actual spacing between terraces in different parts of a field is given in Table 7.4.

TABLE 7.4. Actual and recommended Spacing of Terraces on Irregular Land Slopes

Land Slope (%)	Actual Spacing		Recommended Spacing	
	Vertical interval (ft)	Surface distance (ft)	Vertical interval (ft)	Surface distance (ft)
4	5.6	140	4.4	110
6	5.6	93	5.6	93
8	5.6	70	6.8	85

Compare this actual spacing to the recommended spacing on that slope which is also given in the table. The recommended spacing is obtained by substituting the slope in the equation $VI = 0.6S + 2$.

It is noted from the above comparison that the terraces are closer together than recommended in that section of the field with slopes greater than the average and are farther apart than recommended in that section of the field with slopes less than the average. This variation in terrace spacing in different sections of the field results in (1) point row areas between terraces, thus increasing the difficulty of farming the land; (2) increased difficulty of farming between the closely spaced terraces on the steeper slopes; and (3) increased erosion between the widely spaced terraces on the flatter slopes. Also, if a uniform depth of cut is made during the construction of the terraces on irregular topography, the terraces will usually be quite crooked.

PARALLEL TERRACES

Terraces should be made parallel, with long uniform curves whenever practical. This is extremely important, as the trend is to narrower rows and wider equipment. Following are some of the practices that may be used to improve alignment, reduce curvature, decrease the area of point rows, and improve the farmability of a terraced field:

1. Shape or smooth land prior to construction. If the land surface is rough and irregular, terrace layout will be improved if earth from the ridges is used to fill in depressions. This land smoothing will also reduce the concentration of runoff between terraces and will make it easier to farm the land after the terraces are constructed.

    The amount of shaping and smoothing that is done prior to terrace construction depends on the depth of the soil and the amount of money that will be spent. On shallow soils, the extent of land shaping is limited, since much of the topsoil may be used to fill depressions. On these soils, it may be more desirable to make deeper cuts in the terrace channel during construction and use this subsoil to fill the depressions in the field between terraces.

2. Use good judgment in selecting location and number of outlets. In general the greater the number of outlets used, the easier it is to make the terraces parallel. However, if grassed waterways are used as outlets, the additional outlets will take land out of production and interfere with the operation of farm machinery. If underground terrace outlets are used, additional ones will increase the cost.

    The type of soil and the cropping system to be followed must be considered in selecting the outlet location. Consider the following situations:

    a. A farmer grows only row crops and has no need for the forage produced in a grassed waterway. His soil is relatively deep and productive. This man would prefer to have a limited number of outlets and improve the alignment between terraces by a greater variation in cut in the terrace channel. He may consider the use of underground outlets.

b. A livestock farmer can use the forage produced in the waterways. His soil is relatively shallow and the subsoil unproductive. This man would prefer to have additional grassed waterways and improve the alignment between terraces by varying the grade in the terrace channels.

On any field a number of alternatives should be considered in arriving at the number and location of outlets that will provide the best system of terraces.

3. Use narrow ridges with flat slopes as turn strips. If terraces are staked to carry water around a narrow ridge with a slope much less than the rest of the field, there will be sharp turns in the terraces with large areas of point rows between them. Whenever possible the system should be planned so the terraces will drain away from the ridges. The ridge can then be used as a roadway or turn strip.

4. Plan fence and outlet locations to facilitate farming the terraces. Terraces are more farmable if they approach a fence or an outlet at a right angle. In some cases terraces can be made parallel to fences or property lines; this will eliminate odd areas and point rows.

5. Shift the location of terraces. To reduce the curvature of terraces and make them more nearly parallel it is necessary to shift sections of terraces up or down the slope. The location of graded terraces can be shifted by varying the depth of cut in the terrace channel, varying the grade along the terrace, and by a combination of these two.

In general, varying the depth of cut is more effective in moving short sections of a terrace up or down the slope to reduce the sharpness of curvature. Varying the grade along the terrace is a method used to shift longer sections of a terrace up or down the slope to make terraces more nearly parallel.

The dotted line in Figure 7.14 represents the location of terraces staked using a constant grade throughout their length. The solid line represents the location of the improved terraces. The curvature of the top terrace is reduced by varying the depth of cut. The depth of cut is greater in those sections G where the terrace is above the original location and less in those sections L where the terrace is below the original location, in order to maintain grade in the terrace channel. The lower section of the second terrace up to point A can be made parallel to the top terrace by varying the depth of cut. In the upper section of this second terrace, $A-B$, it will be necessary to vary both the grade and depth of cut. By increasing the grade, the terrace can be moved to the location indicated by the dashed line, $A-C$. This section is still not exactly parallel to the top terrace and

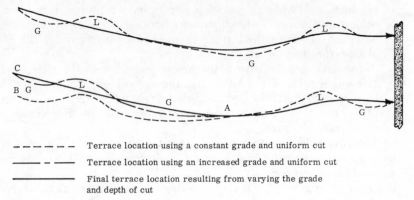

Figure 7.14. *Improving terrace alignment and curvature by varying the depth of cut and the grade.*

it still has one rather sharp curve. The curvature can be reduced and the terrace made parallel to the top terrace by varying the depth of cut. The final location is indicated by the solid line.

When it is necessary to vary the depth of cut, earth is moved from the sections of greater cut to those of lesser cut to maintain the terrace cross section. The allowable range in the depth of cut will vary with the type of soil, the type of equipment used in constructing the terrace, and the amount that the individual is willing to spend.

The extent to which the curvature of the terraces and the point row area between them can be reduced will vary with the topography, the type of soil, the type of equipment to be used in construction, and the amount of money to be spent on construction. The greatest improvement can be made on fields which have relatively uniform topography, moderate slopes, and a deep permeable soil. However, some improvement can be made on any field; the amount will be influenced by the crops to be grown in the field. Most farmers would be willing to spend more to have uniformly spaced terraces with little curvature in a field to be farmed to row crop than they would in a field of small grain or meadow.

Whenever possible, uniform spacing between terraces should be obtained by varying the grade, as there will be no extra cost involved in building them. Whenever the depth of cut is varied, the cost will be increased because of the additional earth movement.

One additional technique that can be used to shift the location of level terraces to make them parallel is to split the terraces into segments that are on different levels. The terrace will be continuous, with the different segments separated by channel blocks to prevent water from flowing to that segment of the terrace on a lower level.

This technique makes it possible to shift a terrace away from the true contour in order to make it parallel to another terrace. It also prevents the excessive earthmoving that would be necessary to bring the entire length of terrace to the same level. The channel blocks should be made with flat side slopes so that they can be farmed over.

Good planning and layout cannot be overemphasized. The additional time that may be required to make the best possible layout will be repaid many times by the reduction in cost of construction.

Use of a Map. A topographic map with a 5-foot contour interval or a contour interval equal to the vertical interval between terraces, if less than 5 feet, will be quite helpful in laying out parallel terraces. If a topographic map is not available, constant grade terraces may be staked and a map made showing their location and elevation. This map can then be used to plan the layout of the parallel terraces. This procedure is discussed in detail in Beasley (1).

The advantage of using a topographic map or a map of constant grade terraces is that all the terraces on the field can be seen at a glance, and adjustments in terrace location can be made much faster than would be possible by restaking the terraces in the field. Also, several tentative systems may be laid out on the map for comparison much faster than this could be done in the field.

The following procedure is suggested:

1. Plan the best possible layout on the map. The slope in different parts of the field should be determined and the terrace spacing selected. The fact that the slope of the entire area can be seen on the map will make it relatively easy to determine which terrace, or section of terrace, would be the best key terrace for that part of the area. Select these key terraces, or sections of terraces, and lay out the others as nearly parallel to them as possible keeping within the limits of grade and depth of cut.

   The layout of a terrace plan on a topographic map is illustrated in Figure 7.15. The average land slope in the upper part of the field is approximately 4 percent, and a terrace spacing of 120 feet is selected. This will be the width required for eight trips with six-row equipment with 30-inch row width. The land slope in the lower part of the field averages 6 percent, so a terrace spacing of 90 feet is selected which is six trips with six-row equipment.

   It would appear that the three upper terraces can be made parallel. The upper half of terrace 3 should be the key. The grades in the upper half of terraces 1 and 2 will be greater than the grade in the key terrace but will not exceed the permissible grade. If the upper end of either terrace 1 or 2 had been selected as the key, the grade in

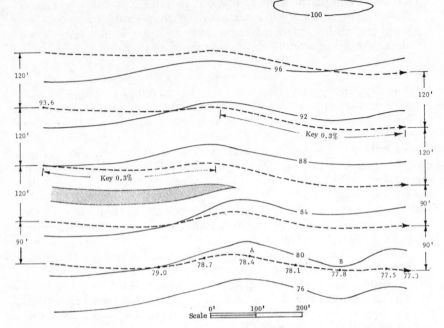

Figure 7.15. Planning a parallel terrace system on a map.

the upper end of terrace 3 would have been too flat or the direction of grade reversed.

The lower half of terrace 2 should be the key for this section. A slight increase in grade in terrace 1 and a slight decrease in grade in terrace 3 will be required to make them parallel to terrace 2. It appears impractical to make terrace 4 parallel to terrace 3.

Whenever it is necessary to vary the spacing between terraces, it should be increased or decreased in intervals equal to or in multiples of the width of equipment that is to be used in farming. In Figure 7.15 the spacing between terraces 3 and 4 is increased from 90 feet at the lower end to 120 feet at the upper end. The resulting point row area is 30 feet wide, which is the width required for one round with the six-row equipment. The change from one spacing to the other should be made in as short a distance as possible so that the point row area will be roughly rectangular in shape rather than a long tapered triangle.

Terrace 5 can be made parallel to terrace 4.

In planning the terrace layout on the map, care must be taken that neither the variation in grade nor the depth of cut will be excessive. It is possible to estimate the approximate grade and depth of cut from the map. For example, in Figure 7.15, the steepest grade appears to be in the upper 300 feet of terrace 2. The elevation at the upper end of the terrace is 93.6 feet. The elevation 300 feet from this point is 92.0—a drop of 1.6 feet in 300 feet. This grade of 0.53 percent is not excessive in this section of the terrace.

The maximum variation in cut appears to be on terrace 5. Assume that a normal cut of 1 foot is made at the point where the terrace crosses the 80-foot contour line. The elevation of the channel of the completed terrace at this point would be 79 feet. The figures along the terrace represent the elevation of the channel, assuming that a 0.3 percent grade is used in the terrace. The least cut will be made at *A* where the elevation of the ground surface is 78.5 feet. The channel elevation is 78.4 feet. A cut of 0.1 foot will be required. The greatest cut will be at *B*. The elevation of the ground surface is 79.6 feet. The elevation of the channel is 77.8 feet. The cut will be 1.8 feet.

2. Transfer the location of one terrace from the map to the field. After the parallel terraces have been planned on the map, one of these must be staked in the same relative position in the field. Reference points must be established in the field and on the map so that this can be done. This terrace must be staked with smooth, uniform curves that will be convenient to farm. It is used as a guide in staking the other terraces. Stakes on parallel terraces with an appreciable variation in depth of cut are usually set at 50-foot intervals.

3. Use the step-offset method for staking the curves. This is a simple method to assure that smooth, uniform curves are staked out. Refer to Figure 7.16 while studying this procedure. Assume that a uniform curve is to be laid out to connect two straight sections of terrace. Step 50 feet from *PC*, the point where the curve begins, in line with the straight section of the terrace. Step sideways a distance equal to one-half the estimated offset and set stake 1. Step out 50 feet in line

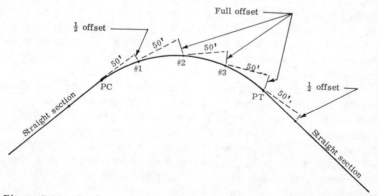

Figure 7.16. Laying out a uniform curve by the step-offset method.

with the two previous stakes set, *PC* and stake 1. Step sideways a distance equal to the estimated offset and set stake 2. Repeat this process until *PT*, the point where the curve ends, is reached using the full offset for each stake. If the curve does not join the straight section at *PT* with the first estimated offset, another offset distance must be selected and the curve restaked. With a little experience it will prove easy to select close to the correct offset on the first trial.

This same procedure can be used to stake out a series of curves not connected to straight sections to assure that there will be no abrupt change in curvature along the terrace.

A curve with a 25-foot offset in 50 feet will be about as sharp as can be farmed with modern machinery. This curve will have a radius of about 100 feet. To determine the approximate radius, divide 2,500 by the offset distance.

$$\text{Radius (ft)} = \frac{2,500}{25} = 100$$

Keep in mind, when selecting the radius of curvature or the offset distance, that terraces staked out or crop rows planted on the inside of the curve will have a smaller radius and will be more difficult to operate machinery around.

4. Stake terraces so they are parallel. After a terrace has been staked, adjacent terraces can be staked parallel to it at the desired spacing. This can be accomplished by two men using a tape or line of the desired length. One man walks along the terrace line already staked, holds one end of the line, and checks to determine if the line is perpendicular to his terrace. The second man holds the other end of the line and moves to the first stake on the adjacent terrace, point 1 in Figure 7.17. The man on the key terrace must be sure that the distance between the terraces is measured perpendicular to his. A positive check to determine if this is the case is made by having the second man hold his end of the line on the stake just set. The man on the key terrace moves forward and back along the terrace. If the measured distance is perpendicular, he will swing away from his terrace in the direction of the second terrace on each side of his measuring position.

The second man steps 50 feet, keeping a taut line, and the other man walks along the key terrace, keeping perpendicular to him. The man on the key terrace will go less than 50 feet in staking a terrace on an outside curve and more than 50 feet in staking a terrace on an inside curve. The man on the second terrace sets the stake, being sure that the line is taut, and the man on the key terrace checks to determine if the measurement is perpendicular to his. The process is repeated for the remaining stakes.

Lay out all terraces with the spacings as indicated on the map.

Field Survey. The layout of a parallel terrace system in the field is more or less a "cut and try" procedure. It is usually necessary to adjust the location of the terrace lines several times to obtain the best system. The amount of adjustment often depends upon the experience of the layout men and the topography of the field.

1. Stake out the top terrace. In staking the top terrace, follow the procedure given previously for nonparallel terraces. The location of the top terrace will probably be changed somewhat later if it is not selected as the key terrace.
2. Space the remaining terraces. Determine the average slope of the area that will drain into each of the remaining terraces. Select the spacing for each terrace. Mark the location of each with a single stake. These stakes should be set where the land slope is about average and near the center of the field if possible.
3. Select the key terrace. This terrace, or section of terrace, should be so located that several terraces can be made parallel to it. The location of the outlet and the topography of the field must be considered in selecting the key terrace. The following points will be helpful in selecting the key. The grade in a parallel terrace below the key terrace will be greater than that in the key if the slope of the land increases as the outlet end of the terrace is approached (Fig. 7.18[a]). The grade in a parallel terrace below the key terrace will be less than that in the key if the slope of the land is less toward the outlet end of the terrace. If the variation in land slope is great enough, it may result in the terrace grading in the opposite direction (Fig. 7.18[b]). On long, irregular slopes it may be necessary to divide the terraces into groups and to select a key for each.
4. Stake out the key terrace. Set stakes at 50-foot intervals using a desirable grade for the soil type.

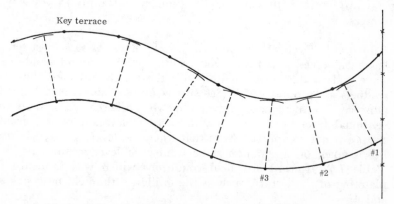

Figure 7.17. Staking a terrace parallel to a key terrace.

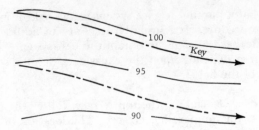

(a) Grade in parallel terrace below the key terrace increases

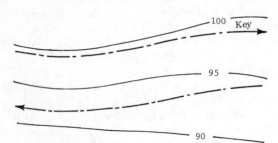

(b) Grade in parallel terrace below the key terrace decreases
(in this case, the grade is reversed)

Figure 7.18. Effect of topography on terrace grades.

5. Adjust the key terrace. Adjust the stakes on the key terrace slightly where necessary to reduce the curvature of sharp or irregular curves.
6. Stake other terraces parallel to the key terraces. Use the same procedure as discussed under Use of a Map.

Computing Terrace Grades and Depth of Cut. The following procedure applies to broad-base graded terraces with a grassed outlet. The same general procedure, with some modification, would apply to other types of terraces.

1. Run profile levels along the terrace stake line. As soon as a terrace has been staked out, take a rod reading on the ground beside each stake and in the outlet in line with the terrace. Record these rod readings as "ground rod." (See survey notes, Table 7.5, column 2.) Observe that these notes are of somewhat different form than conventional level notes and that the actual elevation of the ground surface is not computed. Note that when the level is moved, a reading is taken on a common stake from each level setup (stake 8 in Table 7.5). Record any field conditions which will influence the planning of the terrace, such as the gullies in the field near stakes 9 and 15.

2. Plot a profile of the ground surface along each terrace. To determine the grade in each section of the terrace that will result in the least amount of work in construction, the survey notes should be plotted and a profile of the ground surface drawn (Fig. 7.19). Note that rod readings are plotted rather than elevations and that rod readings are numbered down from the top of the profile sheet. If more than one level setup is made, the profile is plotted as if all readings had been taken from the first one.

In the survey notes (Table 7.5) observe that the second level setup was 2 feet lower in elevation than the previous one. This can be determined by noting the rod reading on stake 8 from each setup. In plotting the profile, 2 feet is added to each rod reading taken from the second setup.

This profile represents the elevation of the ground surface along the terrace. Draw a line along the profile at such grade and elevation that the amount of earth above the line will be approximately equal to the amount below (line A in Fig. 7.19). Plan so the distance that the earth must be moved will be minimal. If additional earth is needed to fill gullies in the field, such as in the vicinity of stakes 9 and 15, then the line should be drawn with additional earth above it in the vicinity of the gullies. Where the profile is above this line, a greater depth of cut will be required than where it is below. The

TABLE 7.5. Survey Notes for Terrace 1

Stake[a]	Ground Rod	Grade Rod	Cut (ft)	Desirable Cut	Excess Cut	Deficient Cut
0	1.9	2.5	0.6	0.6		
1	1.9	2.7	0.8	0.8		
2	2.0	2.8	0.8	0.8		
3	2.8	3.6	0.8	0.8		
4	3.0	3.8	0.8	0.8		
5	3.5	4.3	0.8	0.8		
6	3.8	4.6	0.8	0.8		
7	3.7	4.8	1.1	0.8	0.3	
8	4.1	5.0	0.9	0.8	0.1	
TP						
8	2.1	3.0	0.9	0.8		
9 gully	2.7	3.2	0.5	0.9		0.4
10	2.6	3.4	0.8	0.9		0.1
11	2.5	3.6	1.1	0.9	0.2	
12	2.5	3.7	1.2	0.9	0.3	
13	2.9	3.9	1.0	0.9	0.1	
14	3.2	4.1	0.9	0.9		
15 gully	3.5	4.3	0.8	0.9		0.1
16	3.5	4.5	1.0	1.0		
17	3.4	4.7	1.3	1.0	0.3	
18	3.9	4.9	1.0	1.0		
19	4.4	5.1	0.7	1.0		0.3
19 + 25 outlet	5.2	5.2	0	0		

[a] Stakes are set at 50-foot intervals

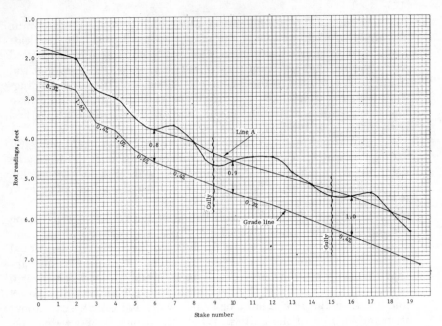

Figure 7.19. Profile of ground surface and proposed grade line for terrace 1.

grade of this line must fall within the allowable limits for the different sections of the terrace (Table 7.2).

3. Compute the depth of cut required at each stake. At a certain depth of cut the earth taken from the channel will be equal to the earth required in the ridge. This depth of cut will vary with the shape of the terrace cross section, the height of the terrace ridge, and the land slope (8). The depth of cut for a V-shaped channel with the cutslope equal in length to the backslope can be computed by the equation:

$$\text{Depth of cut} = \tfrac{1}{2}(h + SW/100)$$

where $h =$ height of the terrace, in feet

$S =$ land slope, in feet per 100 feet, percent slope

$W =$ length of the front slope, in feet

Let us assume that the depth of cut required in this example is 0.8 foot for the upper 400 feet of the terrace, 0.9 foot from 400 to 800 feet, and 1.0 foot for the remainder.

Draw a line on the profile below line A at a distance equal to the depth of cut required. This is designated "grade line" in Figure 7.19.

The grade line represents the location of the bottom of the

completed terrace channel. The "grade rod" gives the rod readings that would have been obtained had it been possible to place the rod at the elevation of the bottom of the completed terrace. These values may be taken directly from the profile; however, it is usually easier to compute them by taking the grade rod reading at the upper end of the terrace and adding the proper grade for each succeeding stake. Record these values to the nearest tenth of a foot in column 3 of your survey notes, as in Table 7.5.

Compute the depth of cut at each stake by subtracting the ground rod from the grade rod and record in column 4.

The grade rod reading at the outlet should be approximately the same as the rod reading taken there. The terrace can be cut into the outlet either upslope or downslope from the point at which the rod reading was taken, if necessary to obtain the proper grade between the end of the terrace and the outlet.

Columns 5, 6, and 7 of Table 7.5 give the desired cut and the excess or deficiency. It is noted that there is an excess of earth between stakes 11 and 13. This excess was planned and will be used to fill the gullies in the field near stakes 9 and 15.

4. Mark the stakes for construction. The depth of cut to be made at each stake should be marked on that stake for use during construction.

CONSTRUCTION

Terraces must be properly constructed if they are to control erosion and not interfere unnecessarily with farming operations. Some of the following discussion of the construction of broad-base and steep-backslope terraces may also be applied to other types. Construction includes the necessary staking, earthmoving in building the terrace, and shaping and smoothing between terraces.

STAKING

Nonparallel Broad-Base Terrace. In constructing nonparallel broad-base terraces, the earth is moved laterally from the channel to the ridge. The earth excavated is sufficient to construct the ridge to the desired height. Since there is little variation in depth of cut required, the stakes set during terrace layout are usually adequate for construction stakes. These indicate the location of the intersection of the front slope of the terrace and the bottom of the terrace channel (Fig. 7.20[a]).

Parallel Broad-Base Terrace. In constructing parallel broad-base terraces, the depth of cut in the terrace channel will vary and earth must be moved longitudinally as well as laterally. The depth of cut to be made at each stake should be marked on the stake (Figs. 7.20[b] and [c]).

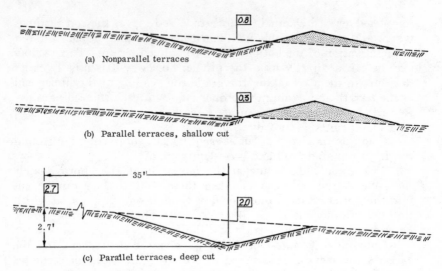

(a) Nonparallel terraces

(b) Parallel terraces, shallow cut

(c) Parallel terraces, deep cut

Figure 7.20.   Construction stakes for broad-base terraces.

Where the deeper cuts are made, it is advisable to set additional reference stakes from which to determine the required depth of cut and terrace location if the channel stake is destroyed during construction. These stakes are usually offset a given distance from the channel stakes, and the difference in elevation between the ground surface at the stake and the bottom of the terrace channel is marked on the stake (Fig. 7.20[c]).

In construction, earth is moved from the areas of deeper cuts, where little or no earth is required in the ridge, to the areas of shallow cut, where additional earth is required in the ridge.

Steep-Backslope Terrace. In constructing steep-backslope terraces, the earth may be taken from both the upper and lower sides of the ridge (Fig. 7.21[a]). It may all be taken from the lower side of the ridge (Fig. 7.21[b]). Earth may be taken from field ridges between terraces (Fig. 7.21[c]) and used to make fills across depressions (Fig. 7.21[d]).

The depth of cut at each channel stake should be marked on the stake. Where an appreciable amount of earth is to be taken from the lower side of the terrace or an appreciable fill is made across a depression, it is advisable to set a stake marking the location of the toe of the backslope (Fig. 7.21[b], [c], and [d]).

EQUIPMENT

Balanced-cross-section terraces can be built with a wide variety of equipment: bulldozers, motor patrols, elevating terracers, plows, graders, scrapers, and wheel tractor hauling units are commonly used. Bulldozers are efficient for building cut-and-fill terraces if the distance that the earth must be moved does not exceed 150 feet. For moving earth greater distances, scrapers pulled by wheel tractors or wheel tractor hauling units are more efficient.

Probably the most efficient combination of equipment for cut-and-fill terraces in general would be a scraper to haul the earth that has to be moved an appreciable distance and a bulldozer to move the earth laterally from the channel to the ridge. The techniques used by equipment operators to move the earth will vary, but in all cases it is absolutely necessary that the operator understand the importance of making the proper depth of cut and finishing the terrace to the desired cross section.

In some cases where unfavorable subsoil is exposed during construction of the terraces, it may be desirable to stockpile topsoil and spread it over the exposed subsoil areas.

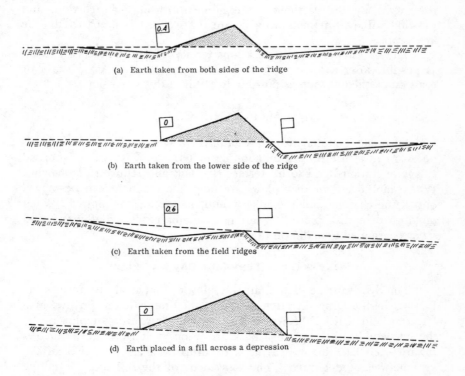

(a) Earth taken from both sides of the ridge

(b) Earth taken from the lower side of the ridge

(c) Earth taken from the field ridges

(d) Earth placed in a fill across a depression

Figure 7.21. Construction stakes for steep-backslope terraces.

SHAPING AND SMOOTHING BETWEEN TERRACES

Shaping and smoothing the land surface between terraces will improve the system. If the land surface between terraces is left irregular, water accumulates in the low places, breaks over the crop rows, and forms silt bars in the terrace channel below. These silt bars reduce the channel capacity and cause wet areas in the terrace channel. Smoothing the land surface reduces the erosion between terraces, conserves moisture, provides for better farming by making it possible to use modern equipment more efficiently, and results in more uniform crop growth.

Larger draws and major depressions between terraces should be filled either prior to or during construction by the equipment used in building the terraces. After the terraces are constructed, the land between should be plowed and smoothed to eliminate minor surface irregularities. The smoothing operation can be done best with some type of land leveler or land plane.

Immediately after the construction of the terraces is completed, the entire area should be fertilized and plowed. The ground will be compacted by the earthmoving equipment during construction, and plowing will permit mellowing before the seedbed is prepared for the first crop.

The backslopes of steep-backslope terraces should be seeded as soon as possible after construction. It is important to get a good stand of grass established as soon as possible to stabilize this steep slope.

MAINTENANCE

Adequate maintenance to retain the ridge height and shape is necessary to assure proper functioning of terraces. Erosion and normal tillage operations tend to fill terrace channels and reduce ridge heights. Where moldboard or disk plows are used, the sediment can be moved out of the channel and the ridge rebuilt as the field is plowed. When moldboard or disk plows are not used, motor graders, bulldozers, and .scrapers are used to maintain the terraces.

NONPARALLEL TERRACES, ONE-WAY PLOW

The disadvantage of the one-way plow is that it always throws the furrows in the same direction. This means that half the furrows in a terraced field will be thrown uphill and half downhill. In plowing by the one-land method, place a backfurrow on each ridge, turn furrows toward each backfurrow leaving a deadfurrow near the center of the area between the terraces. The movement of the soil away from the center of the terrace interval will in time result in a low, poorly drained area that will reduce production. Also, the upper part of the terrace interval will gradually become steeper and the lower part flatter, result-

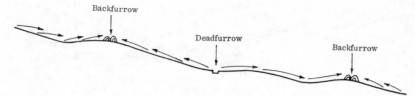

Figure 7.22. Plowing nonparallel terraces with a one-way plow, one-land method.

ing in the land becoming "benched." The terrace channel capacity will gradually be reduced by this method of plowing (Fig. 7.22). The channel capacity can be increased by the two-land method: place a backfurrow on each ridge and upslope about 25 feet from each channel; turn furrows toward each backfurrow leaving deadfurrows as shown in Figure 7.23.

It is sometimes recommended that a backfurrow be placed in the deadfurrow from the previous plowing to reverse the movement of the soil within the terrace interval. This is seldom if ever done on non-parallel terraces because of the inconvenience of plowing the double set of point row areas that will result.

PARALLEL TERRACES, ONE-WAY PLOW

If terraces are made parallel, it is possible to maintain a satisfactory cross section and land slope between terraces with a one-way plow. A backfurrow is placed on each terrace ridge leaving a deadfurrow between the terraces (Fig. 7.24). The next time the field is plowed, a backfurrow is placed on each terrace ridge and on the deadfurrow of the previous plowing. This leaves a deadfurrow in each terrace channel and at the toe of the backslope of each terrace. The terrace channel capacity can be maintained by this method, and also the interval between terraces does not become benched since the soil is moved upslope by one plowing and downslope by the next plowing.

The same general procedure can be used in plowing steep-backslope parallel terraces. The first-year soil is thrown upslope to the toe of the steep backslope and downslope to the terrace ridge. This land is plowed, leaving a deadfurrow in the center of the terrace interval. The next time the field is plowed a backfurrow is placed on the previous dead-furrow.

Figure 7.23. Plowing nonparallel terraces with a one-way plow, two-land method.

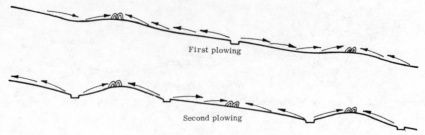

Figure 7.24. Plowing parallel terraces with a one-way plow.

PARALLEL OR NONPARALLEL TERRACES, TWO-WAY PLOW

Either parallel or nonparallel terraces can be adequately main-tained with a two-way plow. If additional channel capacity is needed, the frontslope of each terrace is plowed, turning the furrow to the ridge. In the remaining area, including the point row areas, the furrows are turned upslope, leaving a deadfurrow in the terrace channel (Fig. 7.25[a]). With frequent plowings the terraces may become too large if they are plowed by the above method each time. In this case, all furrows should be turned upslope (Fig. 7.25[b]). In nonparallel terraces it will be neces-sary to plow out the point row areas above each terrace so that the terrace channel and ridge will be plowed with the furrows parallel to the ter-races.

PROBLEMS

7.1. Plan broad-base terraces on Figure P7.1, using transparent paper as an overlay. Use a value of $X = 0.6$ and $Y = 2.0$ in the terrace-spacing formula.

a. Assume the location of the outlet to be at A. (1) Lay out terraces using a constant grade of 0.4 percent. (2) Lay out terraces as nearly parallel as possible by varying the grade and depth of cut. Limit the depth of cut in the terrace channel to less than 3 feet. Keep the grade within the limits specified in the text and in Table 7.2 for a soil of low permeability.

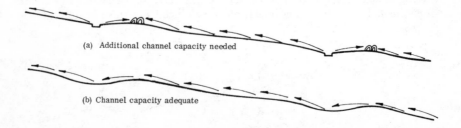

(a) Additional channel capacity needed

(b) Channel capacity adequate

Figure 7.25. Plowing terraces with a two-way plow.

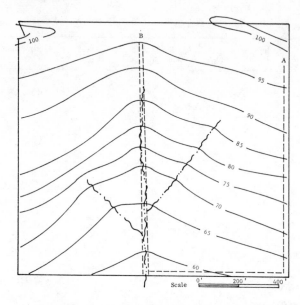

Figure P7.1. Map to be used in planning a terrace system.

b. Assume the location of the outlet to be at *B* and plan the terraces as suggested in (1) and (2) above.

7.2. The terraces indicated by the dotted line on Figure P7.2 have been laid out using a constant grade. The ground surface elevation is noted at each stake. Proposed parallel terraces are indicated by the solid lines.

    a. Plot a profile of the ground surface along the constant grade terraces.

    b. Plot a profile of the ground surface along the proposed parallel terraces on the same graph.

    c. Indicate the depth of cut or height of fill required along the proposed

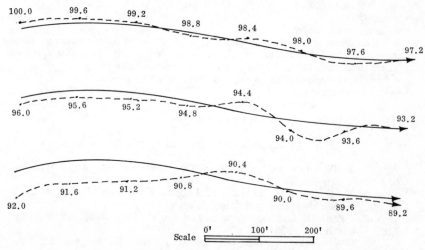

Figure P7.2. Map of constant grade and parallel terraces for study.

TABLE P7.1. Survey Notes for Problem 7.3

Stake	Ground Rod	Grade Rod	Cut	Desirable Cut	Excess Cut	Deficient Cut
0	3.7					
1	3.8					
2	3.9					
3	4.2					
4	4.3					
5	4.3					
6	4.8					
7	5.3					
8	5.4					
9	5.2					
10	5.7					
11	6.4					
12	6.0					
TP						
12	4.5					
13	4.1					
14	4.7					
15	5.1					
16	5.1					
17	5.1					
18	5.1					
19	5.4					
20	6.0					
20 + 20 outlet	6.8					

parallel terrace lines to keep the grade within the limits specified in the text and in Table 7.2 for a soil of low permeability.

7.3. The rod readings in Table P7.1 were taken beside the stakes on a parallel terrace line. The terrace is to be V shaped with the cutslope, frontslope, and backslope each 15 feet long. The land slope is 6 percent. The height of terrace for each 200-foot section starting at the upper end is 0.8, 0.9, 1.0, 1.1, and 1.2 feet respectively.

a. Plot the profile.

b. Determine the terrace grades that will result in a minimum of earth movement during construction. Keep the grades within the limits specified in the text and Table 7.2 for a soil of low permeability.

c. Make survey notes using the information given in Table P7.1 and provide the missing data.

REFERENCES

CLASSIFICATION OF TERRACES

1. Beasley, R. P. A new method of terracing. Univ. Mo. Agr. Expt. Sta. Bull. 699 (rev.), July 1963.
2. Buchta, H. G., Broberg, D. E., and Liggett, F. E. Flat channel terraces. Trans. Am. Soc. Agr. Eng., vol. 9, no. 4, 1966.
3. Jacobson, Paul. Remaking the surface of the earth. Trans. Am. Soc. Agr. Eng., vol. 9, no. 4, 1966.
4. ———. New developments in land-terrace systems. Trans. Am. Soc. Agr. Eng., vol. 9, no. 4, 1966.

5. Nichols, M. L., and Smith, D. D. Progress in erosion control over the past 50 years. *Agr. Eng.*, vol. 38, no. 6, June 1957.

DESIGN AND LAYOUT OF TERRACES

6. Jacobson, Paul. Improvements in bench terraces. Trans. Am. Soc. Agr. Eng., vol. 11, no. 4, 1968.
7. ———. Soil erosion control practices in perspective. *J. Soil Water Conserv.*, vol. 24, no. 4, July–August 1969.
8. Larson, C. L. Geometry of broad-based and grassed-backslope terrace cross sections. Trans. Am. Soc. Agr. Eng., vol. 12, no. 4, 1969.
9. Phillips, R. L., and Beauchamp, K. W. Parallel terraces. USDA Soil Conserv. Serv., January 1966.
10. ———. Design, layout, construction, and maintenance of terraces. American Society of Agricultural Engineers Yearbook, Recommendation R268, 1972.

8 ～ Cross-Slope Channels, Diversions, and Basins

Cross-slope channels, diversions, and
basins are used to dispose of surface runoff
in a manner to prevent excessive erosion.

CROSS-SLOPE CHANNELS

CROSS-SLOPE CHANNELS are wide, shallow channels constructed across the slope to provide drainage and erosion control on nearly flat to gently sloping upland. Farming operations are normally up and down the slope to provide maximum drainage between channels. In plowing with a moldboard or disk plow, the plow is raised in crossing the channel. The channel is plowed lengthwise to maintain the desired cross section. Cross-slope channels may be used on gently sloping land where there are uniform slopes, irregular slopes, or a combination of the two (Figs. 8.1, 8.2, 8.3, and 8.4). In each case some variation in depth of cut and grade in the channel may be used to make the channel more farmable.

DESIGN

Cross Section. The channel may be constructed with a V-shape or with a trapezoidal shape. If it is to be crossed in farming operations, side slopes of ten to one or flatter on V-shaped channels and eight to one or flatter on trapezoidal channels are desirable. If it is not to be crossed, steeper side slopes may be used. The channels should be deep enough to provide drainage from crop rows. The bottom should be at least 0.3 foot below the bottom of the furrow between rows or 0.6 foot below the adjacent field surface. In constructing the channel, the excavated earth may be used either to fill depressions in the field between channels or to build a ridge on the downslope side of the channel (Fig. 8.5).

Spacing. The spacing between channels should be measured at the widest

170

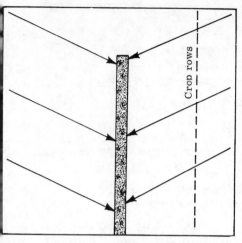

Figure 8.1. Cross-slope channels on uniform, gently sloping land discharging into a grassed waterway.

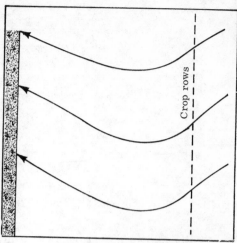

Figure 8.2. Cross-slope channels on irregularly sloping land, discharging into a grassed waterway.

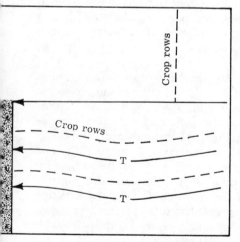

Figure 8.3. Cross-slope channel to intercept runoff and provide drainage for flatland. Terraces control erosion on steeper land below.

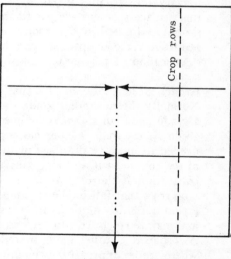

Figure 8.4. Use of a drainage ditch as an outlet for cross-slope channels on flat slopes. The drainage ditch is made deep enough to provide the necessary grade in the cross-slope channels.

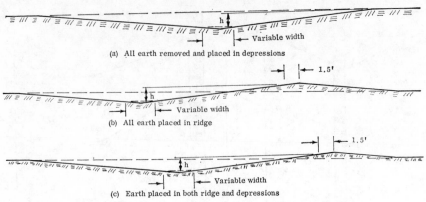

Figure 8.5.   Typical cross-slope channel cross sections.

point and in most cases in the direction of the crop rows. The follow-ing maximum spacings have been found satisfactory in zone 3 of Figure 7.8: 1,320 feet for land slopes up to 0.3 percent, 660 feet for land slopes between 0.3 and 0.6 percent, 440 feet for land slopes between 0.7 and 1.0 percent, and 330 feet for land slopes between 1.1 and 1.5 percent. On very erodible soils, over 1.0 percent slope, the spacing recommended for terraces may be more desirable. In zones other than 3, it may be desirable to vary the spacings given above by approximately the same amount that terrace spacings are varied in different zones. Cross-slope channels should be located to give good drainage on a particular field. This may result in a closer spacing than indicated above. If the land slope is steeper than 1.5 percent terraces should be used.

Grades. The grade in the channels should be within certain limits. Normally the minimum grade should be at least 0.2 percent; however, a grade less than this may be needed in some cases. When the channel grade is less than 0.3 percent, one of the following practices may be needed to improve drainage in the channel: (1) smooth out the bottom of the channel with a drag, cultivator, or grader following planting and cultivation if the crop rows are planted across the channel; (2) plant crop rows that fall in the channel parallel to it to improve drainage; or (3) do not cultivate or plant crops in the channel. Careful construction and maintenance are also necessary to prevent wet spots from developing. A V-shaped channel tends to reduce channel wetness and is desirable where grades approaching the minimum are used.

The maximum grade that can safely be used varies with the erodi-bility of the soil and the amount of runoff expected in the channel. Cross-slope channels are usually constructed on soils which are normally not too erodible. The velocity in a channel that is plowed and planted to row crop or small grain should not exceed 2.0 feet per second for most soils. Velocities can be computed by Manning's formula using an *n*

TABLE 8.1. Depth of Flow (ft) in V-Shaped Cross-Slope Channels with Ten-to-One Side Slopes

| Drainage Area (acres) Land slope (%) | | | Channel Grade (%) | | | | | | |
0–0.4	0.5–0.9	1.0–1.5	0.2	0.3	0.4	0.5	0.6	0.8	1.0
4	3	2	1.0	0.9	0.8	0.8	0.7	0.7	0.6
10	7	5	1.2	1.1	1.0	0.9	0.9	0.8	0.8
20	15	10	1.4	1.3	1.2	1.1	1.0	1.0	0.9
30	22	15	1.6	1.4	1.3	1.2	1.1	1.1	1.0
40	30	20	1.7	1.5	1.4	1.3	1.2	1.2	1.1
50	37	25	1.8	1.6	1.5	1.4	1.3	1.3	
60	45	30	a 1.9	1.7	1.6	1.5	1.4	1.3	
70	52	35	1.9	1.8	1.7	1.6	1.5		
80	60	40	2.0	1.8	1.7	1.7	1.6		
90	67	45	2.1	1.9	1.8	1.7			
	75	50	2.1	2.0	1.8	1.8			
	82	55	2.2	2.0	1.9				
	90	60	2.3	2.1	2.0				

Note: The depth of flow expected with D retardance.

ª Channels above this line may be cultivated and the velocity will not exceed 2.0 feet per second. Channels below this line should not be cultivated. Some grass cover will be required to prevent erosion in the deeper channels. This grass should be mowed frequently so that the flow of water will not be restricted. The velocity based on E retardance will not exceed 3.5 feet per second for any channel listed.

value of 0.03. If the channel is not cultivated, the velocity, computed by using E retardance, should not exceed 3.5 feet per second for most soils.

Tables that will save time in the design of cross-slope channels can be prepared for specific conditions that exist in a given area. Table 8.1 is an example of a table that has been prepared for use in areas having a location factor of 1.0 (Fig. 5.9).

Capacity. Cross-slope channels are usually designed to carry the maximum runoff expected in a five-year to ten-year period. The channels given in Table 8.1 have the capacity to carry the expected runoff from areas having a location factor of 1.0. The depth of flow given in Table 8.1 is expected as a result of D retardance in the channel. This table gives the capacity when the flow is confined within a channel with the indicated depth and side slopes. The capacity of the channel is increased if the ridge is constructed high enough to cause water to flow over the ground surface above the channel. The extent of the increase depends on the height of the ridge and the slope of the land.

The height of the channel should be adequate to contain the depth of flow. The height h (Fig. 8.5), is measured from a point 1.5 feet from the peak of the ridge to a point 1.5 feet from the center of a V-shaped channel or to the bottom of a trapezoidal channel. A channel height greater than the depth of flow given in Table 8.1 may be necessary to provide drainage for crop rows or to provide adequate grade in the channel.

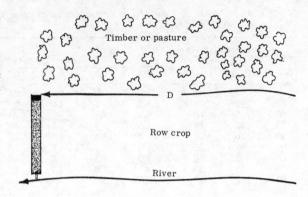

Figure 8.6. Diversion to protect productive bottomland from runoff from upland timber or pasture land.

DIVERSIONS

Diversions are channels constructed across the slope to divert excess water from one location to another where it can be disposed of safely. They may be placed below upland timber or pasture to prevent runoff from these areas from overflowing productive cropland (Fig. 8.6) or near a property line to divert runoff from the other property to a controlled outlet. Debris basins may be required if active gullies are intercepted (Fig. 8.7). A diversion placed near the base of an upland slope protects productive bottomland from runoff from the upland. They may also be used to divert water away from active gully heads to stop their advance, to divert undesirable runoff from farmsteads and water supply reservoirs, to intercept shallow subsurface flow causing wet areas, or to carry runoff from grassed waterways and water storage structures to a stable outlet. Diversions do not make satisfactory substitutes for terraces on land that actually needs terracing.

Figure 8.7. Diversion to protect an upland field from runoff from adjoining property.

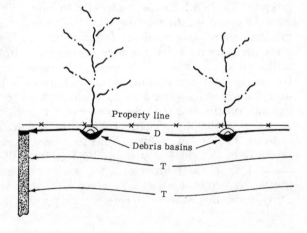

DESIGN

Capacity. The maximum rate of runoff that a diversion is designed to carry is selected by comparing the loss that would occur if the runoff exceeded the capacity with the cost of providing additional capacity. The minimum capacity should not be less than that required to carry a ten-year frequency runoff. However if a diversion carrying water from several hundred acres were located above a farmstead, it would be designed for at least a fifty-year frequency runoff.

Maximum Permissible Velocity. A velocity should be selected which will result in a channel that can be constructed at a minimum cost and can easily be maintained. The use of a higher velocity will result in a smaller channel to carry a given flow; however, the velocity should not be high enough to cause erosion in the channel. Factors to consider in selecting the maximum permissible velocity are the type and condition of the vegetative cover and the erodibility of the soil.

If the diversion is located so that it will be possible to establish and maintain a stand of erosion-resistant grasses in the channel, the section of Table 6.3 for average density grasses may be used in selecting the maximum permissible velocity. A favorable soil is necessary in order to provide a stand of erosion-resistant grasses in the diversion, and in most cases it will be necessary to divert most of the runoff while the vegetation is being established. If it is not possible to divert the runoff from the channel while establishing grass, or an unfavorable soil condition exists in the channel which prevents the establishment and maintenance of a stand of suitable grasses, then the section of Table 6.3 for annual grasses should be used in selecting the maximum permissible velocity.

Channel Grade. The topography at the site where the diversion is to be located will in most cases determine the channel grade. It may be necessary to shift the diversion from the most desirable location in order to obtain a grade which will result in the desired velocity.

Cross Section. A diversion with a trapezoidal-shaped cross section and four-to-one side slopes is most commonly used (Fig. 8.8). The bottom width of the channel and the height of the ridge required to carry a given rate of runoff at or below the maximum permissible velocity can be determined by the procedures given in Chapter 6. However because of the wide range in values for the velocity, the grade in the channel

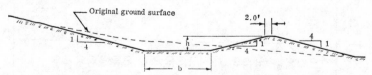

Figure 8.8.   Cross section of a diversion.

TABLE 8.2. Heights (h)* of Diversions with Grass Established and Maintained in Channel

Q, cfs	Width 6 — 0.2	0.4	0.6	0.8	Width 8 — 0.2	0.4	0.6	0.8	Width 10 — 0.2	0.4	0.6	0.8
5	1.6	1.4	1.2		1.5	1.3	1.2	1.1	1.4	1.2	1.1	
10	1.8	1.6	1.3	1.2	1.7	1.5	1.3	1.2	1.6	1.4	1.2	1.1
15	2.0	1.7	1.5	1.4	1.8	1.6	1.4	1.2	1.7	1.5	1.3	1.2
20	2.1	1.8	1.6	1.5	1.9	1.7	1.5	1.4	1.8	1.6	1.4	1.3
25	2.2	1.9	1.7	1.6	2.0	1.8	1.6	1.5	1.9	1.7	1.5	1.4
30	2.3	2.0	1.7	1.6	2.1	1.8	1.6	1.5	2.0	1.7	1.5	1.4
35	2.4	2.1	1.8	1.7	2.2	1.9	1.7	1.6	2.0	1.8	1.6	1.5
40	2.5	2.1	1.9	1.8	2.3	1.9	1.8	1.7	2.1	1.8	1.7	1.6
45	2.6	2.2	2.0	1.8	2.4	2.0	1.9	1.7	2.2	1.9	1.8	1.6
50	2.6	2.3	2.0	1.9	2.4	2.1	1.9	1.8	2.3	2.0	1.8	1.7
60	2.7	2.4	2.1	2.0	2.5	2.2	2.0	1.9	2.4	2.1	1.9	1.8
70	2.8	2.5	2.2	2.0	2.6	2.3	2.1	1.9	2.5	2.2	2.0	1.8
80	2.9	2.6	2.3	2.1	2.7	2.4	2.2	2.0	2.6	2.3	2.1	1.9
90	3.0	2.7	2.4	2.2	2.8	2.5	2.3	2.1	2.7	2.4	2.1	2.0
100	3.0	2.8	2.6	2.3	2.9	2.6	2.3	2.2	2.8	2.5	2.2	2.1
120		2.9	2.7	2.5	3.0	2.7	2.4	2.3	2.9	2.6	2.3	2.2
140		3.0	2.8	2.6		2.8	2.6	2.4	3.0	2.7	2.4	2.3
160			2.9	2.7		2.9	2.7	2.5		2.8	2.6	2.4
180			3.0			3.0	2.8	2.6		2.9	2.7	2.5
200							2.9			3.0	2.8	
220							3.0				2.9	
240											3.0	

Bottom Width (b)** feet; Grade, percent. Circled grade-line values: 1.0, 1.5, 2.0, 2.5, 3.0, 4.0, 5.0.

176

Q, cfs	Bottom Width (b)** feet												Q, cfs
	12				16				20				
	Grade, percent												
	0.2	0.4	0.6	0.8	0.2	0.4	0.6	0.8	0.2	0.4	0.6	0.8	
5	1.4	1.2	1.1		1.3	1.2			1.3	1.2			5
10	1.6	1.4	1.2	1.1	1.5	1.3	1.2	1.1	1.4	1.2	1.1	1.1	10
15	1.7	1.5	1.3	1.2	1.6	1.4	1.3	1.2	1.5	1.3	1.2	1.1	15
20	1.7	1.5	1.3	1.3	1.6	1.4	1.3	1.3	1.5	1.3	1.2	1.2	20
25	1.8	1.6	1.4		1.7	1.5	1.4		1.6	1.4	1.3		25
30	1.9	1.6	1.5	1.4	1.8	1.5	1.4	1.3	1.6	1.5	1.3	1.2	30
35	1.9	1.7	1.5	1.4	1.8	1.6	1.5	1.3	1.7	1.5	1.4	1.3	35
40	2.0	1.7	1.6	1.5	1.9	1.6	1.5	1.4	1.8	1.6	1.4	1.3	40
45	2.1	1.8	1.7	1.5	2.0	1.7	1.6	1.4	1.9	1.6	1.5	1.4	45
50	2.2	1.9	1.7	1.6	2.0	1.8	1.6	1.5	1.9	1.7	1.5	1.4	50
60	2.3	2.0	1.8	1.7	2.1	1.8	1.7	1.6	2.0	1.7	1.6	1.5	60
70	2.4	2.0	1.9	1.7	2.2	1.9	1.8	1.6	2.1	1.8	1.6	1.6	70
80	2.5	2.1	2.0	1.8	2.3	2.0	1.9	1.7	2.1	1.9	1.7	1.6	80
90	2.6	2.2	2.0	1.9	2.4	2.1	1.9	1.8	2.2	1.9	1.8	1.7	90
100	2.7	2.3	2.1	2.0	2.5	2.2	2.0	1.9	2.3	2.0	1.8	1.7	100
120	2.8	2.4	2.2	2.1	2.6	2.3	2.1	2.0	2.5	2.1	1.9	1.8	120
140	3.0	2.6	2.3	2.2	2.8	2.4	2.2	2.1	2.6	2.3	2.0	1.9	140
160		2.7	2.4	2.3	2.9	2.5	2.3	2.2	2.7	2.4	2.1	2.0	160
180		2.8	2.5	2.4	3.0	2.6	2.4	2.3	2.9	2.5	2.2	2.1	180
200		2.9	2.6			2.7	2.5	2.4	3.0	2.6	2.3	2.2	200
220		3.0	2.7			2.8	2.6			2.7	2.4	2.3	220
240			2.8			2.9	2.7			2.8	2.5		240
260			2.9			3.0	2.8			2.9	2.6		260
280										3.0	2.6		280
300											2.7		300

Circled velocity markers (feet per second, on diagonal lines):

- Width 12: 1.0, 1.5, 2.0, 2.5, 3.0, 4.0, 5.0
- Width 16: 1.0, 1.5, 2.0, 2.5, 3.0, 4.0, 5.0
- Width 20: 1.0, 1.5, 2.0, 2.5, 3.0, 4.0, 5.0

* Height, feet, required to carry flow with long grass in channel ("C" retardance), including 0.3-foot freeboard, measured from the bottom of the channel to a point 2 feet off the peak of the ridge.

** Bottom Width required to give desired velocity with short grass in channel ("D" retardance).

(circle) Velocity, in feet per second, to be expected in channel with short grass ("D" retardance).

TABLE 8.3. Heights (h)* of Diversions with Sparse Vegetation in Channel

Q, cfs	Bottom Width (b)** feet — 6				Bottom Width (b)** feet — 8				Bottom Width (b)** feet — 10				Q, cfs
	Grade, percent				Grade, percent				Grade, percent				
	0.2	0.4	0.6	0.8	0.2	0.4	0.6	0.8	0.2	0.4	0.6	0.8	
5	1.3	1.1	1.0	1.0	1.2	1.0	1.0	1.0	1.1	1.0	1.0	1.0	5
10	1.5	1.3	1.1	1.1	1.3	1.2	1.1	1.0	1.3	1.1	1.0	1.0	10
15	1.7	1.4	1.3	1.2	1.5	1.3	1.2	1.0	1.4	1.2	1.1	1.0	15
20	1.8	1.5	1.4	1.3	1.6	1.4	1.3	1.1	1.5	1.3	1.2	1.1	20
25	1.9	1.6	1.5	1.4	1.7	1.5	1.4	1.2	1.6	1.4	1.2	1.2	25
30	2.0	1.6	1.5	1.5	1.8	1.6	1.5	1.3	1.7	1.5	1.3	1.3	30
35	2.1	1.7	1.6		1.9	1.7	1.5	1.4	1.8	1.6	1.3	1.3	35
40	2.2	1.8	1.7		2.0	1.7	1.6		1.9	1.6	1.4		40
45	2.3	1.9	1.7		2.1	1.8			2.0	1.7	1.5		45
50	2.3	2.0			2.2	1.9			2.1	1.8			50
60	2.4				2.3	2.0			2.2	1.9			60
70	2.5				2.4				2.3				70
80	2.7				2.5				2.5				80
90	2.8				2.5				2.6				90
100	2.9				2.6				2.6				100
120	3.1				2.8				2.7				120
140	3.3				3.0				2.8				140

Circled contour values (h) marked within each block: 1.0, 1.5, 2.0, 2.5, 3.0, 3.5

178

Q, cfs	Bottom Width (b)** feet												Q, cfs
	12				16				20				
	Grade, percent												
	0.2	0.4	0.6	0.8	0.2	0.4	0.6	0.8	0.2	0.4	0.6	0.8	
5	1.1	1.0	1.0	1.0	1.0	1.0	1.0	1.0	1.0	1.0	1.0	1.0	5
10	1.3	1.1	1.0	1.0	1.2	1.0	1.0	1.0	1.1	1.0	1.0	1.0	10
15	1.4	1.2	1.1	1.0	1.3	1.1	1.0	1.0	1.2	1.1	1.0	1.0	15
20	1.5	1.3	1.1	1.0	1.4	1.2	1.1	1.0	1.3	1.1	1.0	1.0	20
25	1.6	1.4	1.2	1.1	1.5	1.3	1.1	1.0	1.4	1.2	1.1	1.0	25
30	1.7	1.5	1.3	1.2	1.6	1.3	1.2	1.1	1.4	1.3	1.1	1.0	30
35	1.7	1.5	1.4	1.3	1.6	1.4	1.2	1.2	1.6	1.3	1.2	1.1	35
40	1.8	1.6	1.4	1.3	1.7	1.4	1.3	1.2	1.6	1.4	1.2	1.1	40
45	1.9	1.6	1.5		1.7	1.5	1.4		1.7	1.4	1.3	1.2	45
50	2.0	1.7	1.5		1.8	1.6	1.4		1.7	1.5	1.3	1.2	50
60	2.1	1.8			1.9	1.7	1.5		1.8	1.5	1.4		60
70	2.2	1.9			2.0	1.7	1.6		1.9	1.6	1.5		70
80	2.3				2.1	1.8			1.9	1.7			80
90	2.4				2.2	1.9			2.0	1.7			90
100	2.5				2.3				2.1	1.8			100
120	2.7				2.5				2.3				120
140	2.8				2.6				2.4				140
160					2.7				2.5				160

Velocity circles (feet per second):
Width 12: 1.0, 1.5, 2.0, 2.5, 3.0, 3.5
Width 16: 1.0, 1.5, 2.0, 2.5, 3.0, 3.5
Width 20: 1.0, 1.5, 2.0, 2.5, 3.0, 3.5

* Height, feet, required to carry flow with long grass in channel ("D" retardance), including 0.3-foot freeboard measured from the bottom of channel to a point 2 feet off the peak of the ridge. One foot is considered minimum height.

** Bottom width required to give desired velocity with sparse vegetation in channel ("E" retardance).

◯ Velocity, in feet per second, to be expected in channel with short, sparse vegetation ("E" retardance).

179

and the bottom width and depth of channel that are possible, numerous solutions would be required to determine the one most suitable. Tables have been prepared which can be used to determine rapidly the size channel required for trapezoidal waterways with four-to-one side slopes and channel grades up to 0.8 percent.

Table 8.2 should be used if the diversion is located so that it will be possible to establish and maintain a stand of erosion-resistant grasses in the channel. If it is impossible to divert the runoff from the channel or there are other unfavorable conditions which prevent the establishment and maintenance of a stand of erosion-resistant grass in the channel, Table 8.3 should be used.

The channel widths and velocities given in Table 8.2 are based on D retardance, and those in Table 8.3 are based on E retardance. The heights of the diversions given in Table 8.2 are based on C retardance and include 0.3 foot of freeboard, and those in Table 8.3 are based on D retardance and include 0.3 foot of freeboard.

A number of different sizes of diversions can usually be used for a given situation. The size selected will depend on the slope of the land on which the diversion is constructed, the location of the diversion, the type of equipment to be used in construction, and the cost. It is more difficult to construct a wide, shallow diversion on steeply sloping land. If a diversion is located so that it must be crossed frequently, or if the ridge is to be farmed, a wider channel with less height may be preferred. A size should be selected that will be the easiest and least expensive to construct and maintain with the equipment to be used.

Outlet. Each diversion should have an adequate outlet. The outlet may be a grassed waterway, a stable watercourse, a grade stabilization structure, or other type of stable outlet.

BASINS

Basins are used to store excess surface runoff until it infiltrates into the soil. They have a greater capacity to store water than graded diversions have. Their cross sections may vary widely depending on the location. On flat-to-gentle slopes a cross section similar to the flat-channel terrace may be used. Basins are normally designed to store a ten-year frequency runoff with an emergency spillway at either end to handle larger storms. They are usually built level, and the water infiltrates into the soil. However on slightly less permeable soils they may be constructed with a slight grade and an outlet similar to an underground terrace outlet, or a pipe spillway is used to assist in the removal of water to prevent damage to the crop in the basin.

Basins may be placed near the base of upland slopes to protect productive cropland from upland runoff, or below farmsteads and feedlots to prevent runoff from polluting streams. They may be used to collect and utilize irrigation waste water.

PROBLEMS

8.1. A V-shaped cross-slope channel with ten-to-one side slopes is to be constructed in northeast Missouri on a square 40-acre field. The land slope from north to south is 0.5 percent and from east to west, 0.1 percent. The crop rows will be planted north and south. A depression approximately 2.5 feet deep runs along the west boundary of the field.

    a. Indicate the location of cross-slope channels on this field.

    b. Indicate the grade that you would use in different sections of the channel.

    c. Indicate the depth of channel required at points 330 feet, 660 feet, 990 feet, and 1,320 feet from the east boundary.

    d. Indicate those sections of the channel that could be cultivated and those that should not be cultivated.

8.2. A diversion similar to that shown in Figure 8.6 is to be constructed to prevent the runoff from 50 acres of timber from overflowing the bottom field, which is used to produce high cash value crops and is surface irrigated. The peak rates of runoff at the outlet end of the diversion are 60 cfs for a 10-year frequency storm, 78 cfs for a 25-year frequency storm, and 90 cfs for a 50-year storm. The peak rate of runoff decreases gradually to practically 0 cfs at the upper end of the diversion. The soil on which the channel is to be built is quite resistant to erosion. The runoff from the timber cannot be diverted while grass is being established in the channel. The diversion will discharge into a creek through a pipe spillway, and will not have to be crossed except for maintenance. The land slope at the diversion location is 4 percent. A bulldozer having a 10-foot blade will be used for construction. Give the peak rates of runoff used, the type of vegetation planned in the channel, the grade used, the velocity to be expected, and the width and height of the different sections of the diversion.

8.3. The diversion in Figure 8.7 is to be constructed in central Missouri on a silt loam soil. The land slope immediately below the property line is 8 percent; most of the runoff is concentrated in the two gullies. Runoff from 15 acres of cropland enters the diversion from the gully farthest from the outlet. The peak rates of runoff to be expected from 10-year, 25-year, and 50-year frequency storms is 50 cfs, 65 cfs, and 75 cfs respectively. The runoff from an additional 20 acres of cropland enters the diversion from the gully nearest the outlet. The total peak rates of runoff to be expected from 10-year, 25-year, and 50-year frequency storms are 110 cfs, 145 cfs, and 165 cfs respectively. In order for the diversion to be located near the property line, a grade of 0.4 percent will have to be used. The area between the diversion and the property line will be kept in grass. The field below is usually in a C-W-M-M rotation. The diversion will be constructed with a scraper having an 8-foot cutting blade. Give the peak rates of runoff used, the type of vegetation planned in the channel, the velocity to be expected, and the width and height of the different sections of the diversion.

9 ~ Spillways and Earth Embankments

Structures stabilize the grade in natural or artificial channels, prevent formation and advance of gullies, store water, sediment, and debris, control irrigation and drainage water, and reduce flooding.

GOOD VEGETATIVE PRACTICES, together with proper land use, are indispensable in a sound soil and water management program. There are many instances however when vegetative measures alone are inadequate to handle the concentration of water. Structures are vital in reinforcing or supplementing these practices.

A structure may consist of a combination of one or more of the following parts: a principal spillway, an emergency spillway, and an earth embankment.

PRINCIPAL SPILLWAY

A principal spillway may be a pipe spillway, a drop spillway, or a chute spillway.

PIPE SPILLWAY

A pipe spillway consists of a closed conduit with an inlet designed to cause the conduit to flow full with a relatively low depth of water over the entrance and with all the appurtenances necessary to provide a safe structure. It may be used to discharge water directly from diversions, grassed waterways, or other channels into a main drainageway. It may be constructed with no provision for permanent water storage behind the structure and very little detention storage above the spillway inlet. The outlet end of the conduit is usually placed above the expected water level. The bank of the drainageway is sloped so

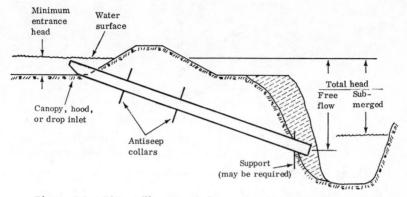

Figure 9.1. Pipe spillway to discharge water into a drainageway.

that the conduit does not protrude into the channel of the drainageway (Fig. 9.1). Pipe spillways are also used in conjunction with earth embankments to carry water from above the embankment to a lower elevation (Fig. 9.2).

If a conduit with the end cut off square is used as the spillway in a structure as in Figure 9.3, the flow lines of the water entering the conduit will be directed downward and it will not flow full at reasonable depths of water over the inlet. The capacity will be the result of head *h* (Fig. 9.3) on the end of the conduit. The conduit would flow full, and a greater capacity would be obtained if a depth of flow over the inlet equal to five to seven times the diameter of the conduit could be obtained. It is seldom practical to impound this much water above the inlet.

The *canopy inlet* (Fig. 9.4) directs the flow lines of the water upward, causing the conduit to fill soon after the end of the conduit is covered with water. The depth of water above the inlet when the conduit flows full is designated the minimum entrance head. With the

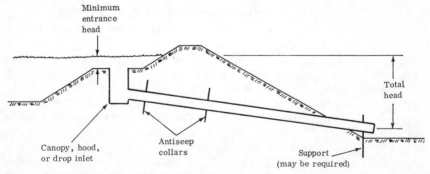

Figure 9.2. Pipe spillway to discharge water impounded behind an embankment.

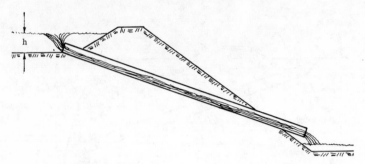

Figure 9.3. Flow conditions in a conduit with a square end.

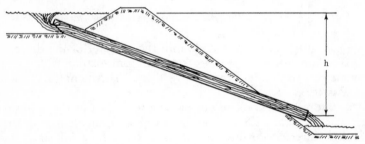

Figure 9.4. Flow conditions in a conduit with a canopy inlet.

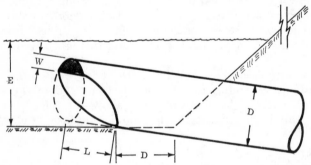

Figure 9.5. The canopy inlet. (See Table 9.1 for dimensions.)

conduit flowing full, the head-causing flow h is the vertical distance from the water surface at the inlet to the center of the outlet end of the conduit, or to the water level if the outlet end is submerged. The canopy inlet is formed by cutting the end of the conduit and welding an end plate of suitable material to the opening. (See Fig. 9.5 and Table 9.1.)

The hood, drop, and morning glory inlets, if constructed to specified dimensions, will also result in full conduit flow at relatively low depths of water over the inlet.

The *hood inlet* is formed by cutting the end of the conduit as in-

TABLE 9.1. Canopy Inlet Dimensions

Slope of Conduit (%)	W^a Plate Width[c]	L Canopy Length[c]	E^b Minimum Entrance Head[c]
0–5	0.2D	0.6D	1.3D
6–15	0.2D	0.8D	1.4D
16–25	0.3D	1.1D	1.5D
26–32	0.35D	1.3D	1.6D

[a] Measured from the valley of the corrugations on corrugated pipe and from the inside surface of smooth pipe.
[b] Minimum entrance head required to cause the conduit to flow full.
[c] D is conduit diameter.

dicated in Figure 9.6 and welding or bolting a baffle of suitable material to the conduit.

The *drop inlet* consists of a vertical riser connected to a conduit (Fig. 9.7). The riser may be made of corrugated or smooth metal pipe or of concrete. The dimensions of the drop inlet and the depth of flow required above the entrance to assure full conduit flow are given in Figure 9.7 and Table 9.2.

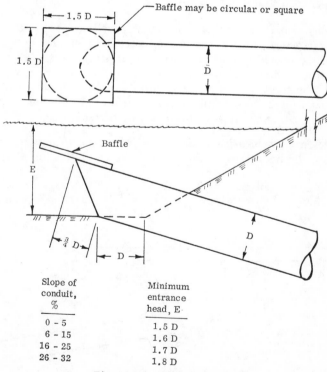

Slope of conduit, %	Minimum entrance head, E
0 – 5	1.5 D
6 – 15	1.6 D
16 – 25	1.7 D
26 – 32	1.8 D

Figure 9.6. The hood inlet.

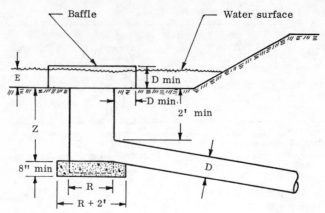

Figure 9.7. The drop inlet. (See Table 9.2 for dimensions.)

The *morning glory inlet* is constructed by making a mound of earth about the conduit to the desired shape of the structure and placing concrete on the earth form. Hand labor is required to form the earth and place the concrete. The dimensions of the morning glory inlet are given in Figure 9.8.

Adaptability. Pipe spillways are particularly well adapted to sites where a good emergency spillway or an appreciable amount of detention storage can be provided, and to controlling overfalls greater than 8 feet. If high rates of flow are to be carried, a suitable emergency spillway cannot be constructed, and there is limited detention storage, a large diameter conduit will be required, which is undesirable and expensive. In this case, a chute or drop spillway may be better suited.

Advantages and Limitations. Compared with the drop and morning glory inlets, the hood and canopy inlets have the following advantages:

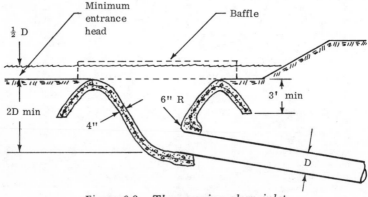

Figure 9.8. The morning glory inlet.

TABLE 9.2. Dimensions for Drop Inlets

Conduit Diameter D (in)	Riser Height Z between 2D and 4D				Riser Height Z 5D or Greater			
	Riser diameter R (in)		Entrance head E (ft)		Riser diameter R (in)		Entrance head E (ft)	
	CMP[a] conduit	WSP[b] conduit	CMP[a] conduit	WSP[b] conduit	CMP[a] conduit	WSP[b] conduit	CMP[a] conduit	WSP[b] conduit
6	12	12	0.4	0.5	12	12	0.4	0.5
8	12	12	0.6	0.8	12	12	0.6	0.8
10	15	14	0.8	1.0	15	14	0.8	1.0
12	18	18	0.9	1.1	15	16	1.0	1.3
14		20		1.2		18		1.4
15	24		1.0		21		1.2	
16		24		1.4		20		1.5
18	30	28	1.3	1.5	24	22	1.5	1.7
20		30		1.7		26		1.8
21	30		1.6		26		1.7	
22		34		1.8		28		2.0
24	36	36	1.7	1.9	30	30	1.9	2.1
26		42		1.9		36		2.0
28		42		2.0		36		2.2
30	48	48	1.9	2.1	42	42	2.1	2.4
36	54	54	2.3	2.5	48	48	2.5	2.8

[a] Corrugated metal pipe
[b] Welded steel pipe

no riser is required; there is less excavation required if the structure is used at the end of waterways or diversions to discharge water into a main drainageway; there is a lower fill over the conduit in structures using an earth embankment; they are simple to fabricate and install; they are low in first cost; they are less likely to be clogged with debris; and livestock and children cannot fall into them and be injured.

Compared to the hood inlet, the canopy inlet has the following advantages: the conduit will fill at a lower entrance head; and since a baffle is not required, it is simpler to fabricate and install and is less expensive.

Canopy and hood inlets require a greater depth of water over the inlet to obtain full conduit flow than do the drop or morning glory inlets. For larger diameter conduits, a large difference in elevation between the conduit inlet and the emergency spillway is required if full conduit flow is to occur before the emergency spillway functions. This disadvantage may be overcome by using an entrance box on the inlet as shown in Figure 9.9.

Design and Construction. The *conduit* for small structures is usually made of corrugated or smooth metal pipe. Reinforced concrete pipe may be used in large structures. The minimum gage of metal for corrugated galvanized metal pipe, corrugated aluminum alloy pipe, and welded steel pipe with various fill heights is given in Table 9.3.

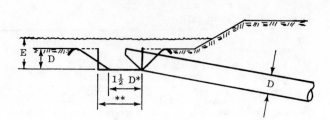

* Minimum diameter for a circular box with sloping sides.

** Box with vertical sides may be either round or square.
   Use minimum dimension of 2D for the side if square or
   2.5D for the diameter if round.

Figure 9.9. Entrance box for canopy or hood inlet structures.

The pipe is placed on a solid foundation throughout its length, and corrugated metal pipe laid with the outside laps of circumferential joints pointing upstream with longitudinal laps on the sides. Where no cradle is provided under the pipe, the foundation is shaped to conform to the bottom of the pipe for at least one-tenth of its diameter. Backfill material is thoroughly compacted on each side of the pipe.

Antiseep collars should have a minimum projection of 1.0 foot beyond the pipe for pipes less than 12 inches in diameter, 1.5 feet for pipes 12–24 inches in diameter, and 2.0 feet for pipes 30 inches in diameter and greater. A sufficient number of antiseep collars should be used to increase the line of seepage by at least 15 percent for morning glory or drop inlets and 10 percent for canopy and hood inlets. The number of collars required may be computed by the following equation:

$$N = \frac{CL}{V}$$

where N = number of collars required
 C = 0.15 for morning glory or drop inlets, 0.10 for canopy or hood inlets
 L = length of saturated zone (total length of pipe minus 20 feet)
 V = total width or height of the collar in feet minus the pipe diameter in feet (maximum V to be 5 feet)

Antiseep collars should not be more than 25 feet apart and spaced with approximately equal distances between them. The first collar on the upstream end should usually be less than one-half a full spacing from the inlet. The last collar should not be closer than a full spacing from the end of the pipe.

An *outlet end support* may be required where the conduit projects some distance beyond the supporting earth or where large diameter conduits are used.

TABLE 9.3. Minimum Pipe Gage

Diameter of Pipe (Inches)	Height of Fill (Feet) Over Pipe														
	Up to 15			15 to 20			20 to 25			25 to 30			30 to 35		
	WSP	CMP	CAP	WSP	CMP	CAP	WSP	CMP	CAP	WSP	CMP	CAP	WSP	CMP	CAP
4	12	-	-	12	-	-	12	-	-	12	-	-	12	-	-
6	12	-	-	12	-	-	12	-	-	12	-	-	12	-	-
8	12	16	16	12	16	16	12	16	16	12	16	16	12	16	16
10	12	16	16	12	16	16	12	16	16	12	16	16	12	16	16
12	12	16	16	12	16	16	12	16	16	12	16	16	12	16	16
14	12	-	-	12	-	-	12	-	-	12	-	-	10	-	-
15	-	16	16	-	16	16	-	16	16	-	16	16	-	16	14
16	12	-	-	12	-	-	10	-	-	10	-	-	10	-	-
18	12	16	16	10	16	16	10	16	16	10	16	14	7	16	12
20	10	-	-	10	-	-	10	-	-	7	-	-	7	-	-
21	-	16	16	-	16	16	-	16	14	-	16	12	-	16	10
22	10	-	-	10	-	-	7	-	-	7	-	-	7	-	-
24	10	14	14	10	14	14	7	14	14	7	14	12	7	14	10
30	7	14	14	7	14	14	7	14	12	3	14	10	3	12	-
36	7	14	12	3	12	12	3	12	10	3	12	-	3	10	-
42	3	12	-	3	12	-	3	12	-	$\frac{1}{4}$"	10	-	$\frac{1}{4}$"	10	-

Note: WSP = welded steel pipe; CMP = corrugated galvanized metal pipe; CAP = corrugated aluminum pipe. Pipe gages and the equivalent thickness in inches are given below:

16 gage—.060 inches   8 gage—.164 inches
14 gage—.075 inches   7 gage—.179 inches
12 gage—.105 inches   3 gage—.239 inches
10 gage—.135 inches

TABLE 9.4. Capacities of Canopy, Hood, Morning Glory, and Drop Inlet Spillways with Corrugated Metal Conduit, cfs

Length of Conduit, Feet	Total Head, Feet										
	4	6	8	10	12	14	16	18	20	22	24
6-inch Diameter Conduit											
20	1.1	1.2	1.4	1.5							
30	1.0	1.1	1.3	1.4	1.6	1.7					
40	0.9	1.0	1.2	1.3	1.5	1.6	1.7	1.8	1.9	2.0	2.1
50	0.8	0.9	1.1	1.2	1.3	1.4	1.5	1.6	1.7	1.8	1.9
60	0.8	0.9	1.0	1.1	1.2	1.3	1.4	1.5	1.6	1.7	1.7
70	0.7	0.8	0.9	1.0	1.1	1.2	1.3	1.4	1.5	1.5	1.6
80	0.7	0.8	0.9	1.0	1.1	1.2	1.2	1.3	1.4	1.5	1.5
90	0.6	0.7	0.8	0.9	1.0	1.1	1.2	1.2	1.3	1.4	1.4
100	0.6	0.7	0.8	0.9	1.0	1.1	1.1	1.2	1.2	1.3	1.4
8-inch Diameter Conduit											
20	2.3	2.6	2.9	3.2							
30	2.1	2.4	2.7	3.0	3.3	3.5					
40	1.9	2.2	2.5	2.8	3.1	3.3	3.5	3.7	3.9	4.1	4.3
50	1.7	2.0	2.3	2.5	2.8	3.0	3.2	3.4	3.6	3.8	3.9
60	1.6	1.8	2.1	2.4	2.6	2.8	3.0	3.2	3.3	3.5	3.7
70	1.5	1.7	2.0	2.2	2.4	2.6	2.8	3.0	3.1	3.3	3.4
80	1.4	1.6	1.9	2.1	2.3	2.5	2.6	2.8	3.0	3.1	3.2
90	1.3	1.5	1.8	2.0	2.2	2.3	2.5	2.7	2.8	2.9	3.1
100	1.3	1.5	1.7	1.9	2.1	2.2	2.4	2.5	2.7	2.8	2.9
10-inch Diameter Conduit											
20	3.9	4.4	5.1	5.6							
30	3.6	4.1	4.7	5.2	5.7	6.1					
40	3.3	3.8	4.4	4.9	5.3	5.7	6.1	6.5	6.9	7.2	7.5
50	3.0	3.5	4.0	4.5	4.9	5.2	5.7	6.0	6.3	6.6	6.9
60	2.8	3.2	3.7	4.2	4.6	4.9	5.3	5.6	5.9	6.2	6.4
70	2.6	3.0	3.5	3.9	4.3	4.6	4.9	5.2	5.5	5.8	6.1
80	2.5	2.9	3.3	3.7	4.1	4.4	4.7	5.0	5.3	5.5	5.8
90	2.3	2.7	3.1	3.5	3.9	4.2	4.5	4.7	5.0	5.2	5.4
100	2.2	2.6	3.0	3.4	3.7	4.0	4.3	4.5	4.8	5.0	5.2
12-inch Diameter Conduit											
20	5.4	6.7	7.6	8.7							
30	5.1	6.3	7.2	8.2	9.0	9.7					
40	4.8	5.9	6.8	7.7	8.4	9.1	9.7	10	11	11	12
50	4.5	5.5	6.3	7.1	7.8	8.4	9.0	9.5	10	10	11
60	4.2	5.1	5.9	6.6	7.2	7.8	8.4	8.9	9.3	9.8	10
70	3.9	4.8	5.6	6.2	6.8	7.4	7.9	8.4	8.8	9.2	9.7
80	3.7	4.6	5.3	5.9	6.5	7.0	7.5	7.9	8.4	8.8	9.2
90	3.5	4.3	5.0	5.6	6.1	6.6	7.1	7.5	7.9	8.3	8.7
100	3.4	4.1	4.8	5.4	5.9	6.3	6.8	7.2	7.6	7.9	8.3
15-inch Diameter Conduit											
20	9.5	11	13	14							
30	9.0	10	12	14	15	16					
40	8.4	10	12	13	15	16	17	18	19	20	21
50	7.8	9.6	11	12	14	15	16	17	17	18	19
60	7.3	9.0	10	12	13	14	15	16	16	17	18
70	8.9	8.5	9.8	11	12	13	14	15	15	16	17
80	6.6	8.0	9.3	10	11	12	13	14	15	15	16
90	6.3	7.7	8.9	10	11	12	13	13	14	15	15
100	6.0	7.4	8.5	9.5	10	11	12	13	13	14	15

TABLE 9.4. *(Continued)*

Length of Conduit, Feet	Total Head, Feet										
	4	6	8	10	12	14	16	18	20	22	24
18-inch Diameter Conduit											
20	14	17	19	23							
30	13	16	18	22	24	25					
40	13	16	18	21	23	24	26	28	29	30	32
50	12	15	17	19	21	23	24	26	27	29	30
60	11	14	16	18	20	21	23	24	26	27	28
70	11	13	15	17	19	20	22	23	24	26	27
80	10	13	15	16	18	19	21	22	23	24	25
90	9.9	12	14	16	17	19	20	21	22	23	24
100	9.6	12	14	15	17	18	19	20	21	23	23
21-inch Diameter Conduit											
20	21	25	28	31							
30	20	24	27	30	33	37					
40	19	23	26	29	32	35	37	39	42	44	46
50	18	21	25	28	30	33	35	37	39	41	43
60	17	20	23	26	29	31	33	35	37	39	41
70	16	19	22	25	27	30	32	33	35	37	39
80	15	19	21	24	26	28	30	32	34	36	37
90	15	18	21	23	25	27	29	31	32	34	36
100	14	17	20	22	24	26	28	30	31	33	34
24-inch Diameter Conduit											
20	27	33	38	43							
30	26	32	37	41	46	50					
40	25	31	36	40	44	48	51	54	57	60	62
50	24	30	34	38	42	45	48	51	54	56	59
60	23	28	32	36	39	43	46	48	51	53	56
70	22	27	31	35	38	41	44	46	49	51	54
80	21	26	30	33	36	39	42	45	47	49	51
90	20	25	28	32	35	38	40	43	45	47	49
100	19	24	27	31	34	36	39	41	44	46	48
30-inch Diameter Conduit											
20	44	56	64	72							
30	43	54	62	69	76	82					
40	42	52	60	67	73	79	84	89	94	99	103
50	41	50	58	64	70	76	81	86	91	95	100
60	39	48	55	61	67	73	78	82	87	91	95
70	37	46	53	59	65	70	75	79	83	88	91
80	36	44	51	57	62	67	72	76	80	84	88
90	35	43	49	55	60	65	69	74	78	81	85
100	34	41	47	53	58	63	67	71	75	79	82
36-inch Diameter Conduit											
20	68	83	96	107							
30	66	80	93	104	115	123.					
40	64	78	90	101	111	119	128	135	143	150	156
50	62	75	87	97	107	115	123	131	138	144	151
60	59	73	84	94	103	111	119	126	133	139	145
70	57	70	81	90	99	107	114	121	127	134	140
80	55	68	78	87	96	103	111	117	124	130	135
90	54	66	76	85	93	100	107	114	120	126	131
100	52	63	73	82	90	97	104	110	116	122	127

191

TABLE 9.5.   Capacities of Canopy, Hood, Morning Glory, and Drop Inlet Spillways with Smooth Metal Conduits, cfs

Length of Conduit, Feet	Total Head, Feet										
	4	6	8	10	12	14	16	18	20	22	24
	6-inch Diameter Conduit										
20	1.7	2.1	2.4	2.6							
30	1.5	1.8	2.1	2.4	2.6	2.8					
40	1.4	1.7	2.0	2.2	2.4	2.6	2.8	2.9	3.1		
50	1.3	1.6	1.8	2.0	2.2	2.4	2.5	2.7	2.9	3.0	3.2
60	1.2	1.5	1.7	1.9	2.1	2.3	2.4	2.6	2.7	2.8	3.0
70	1.1	1.4	1.6	1.8	2.0	2.1	2.3	2.4	2.6	2.7	2.8
80	1.1	1.3	1.5	1.7	1.9	2.0	2.2	2.3	2.4	2.6	2.7
90	1.0	1.3	1.5	1.6	1.8	1.9	2.1	2.2	2.3	2.4	2.6
100	1.0	1.2	1.4	1.6	1.7	1.9	2.0	2.1	2.2	2.3	2.5
	8-inch Diameter Conduit										
20	3.2	3.9	4.5	5.1							
30	3.0	3.6	4.2	4.7	5.1	5.5					
40	2.8	3.4	3.9	4.3	4.8	5.2	5.5	5.9	6.2		
50	2.6	3.2	3.7	4.1	4.5	4.8	5.2	5.5	5.8	6.1	6.3
60	2.4	3.0	3.5	3.9	4.3	4.6	4.9	5.2	5.5	5.8	6.0
70	2.3	2.9	3.3	3.7	4.0	4.4	4.6	5.0	5.2	5.5	5.7
80	2.2	2.7	3.2	3.5	3.9	4.2	4.5	4.7	5.0	5.2	5.5
90	2.1	2.6	3.0	3.4	3.7	4.0	4.3	4.5	4.8	5.0	5.3
100	2.0	2.5	2.9	3.2	3.6	3.9	4.1	4.4	4.6	4.8	5.0
	10-inch Diameter Conduit										
20	5.2	6.4	7.3	8.2							
30	4.9	6.0	6.9	7.7	8.5	9.1					
40	4.6	5.6	6.5	7.3	8.0	8.6	9.2	9.8	10.3		
50	4.3	5.3	6.2	6.9	7.6	8.2	8.7	9.3	9.8	10.3	10.7
60	4.1	5.0	5.9	6.6	7.2	7.8	8.3	8.8	9.3	9.8	10.2
70	4.0	4.8	5.6	6.3	6.9	7.5	8.0	8.5	8.9	9.4	9.8
80	3.8	4.7	5.4	6.1	6.6	7.2	7.7	8.1	8.6	9.0	9.4
90	3.7	4.5	5.2	5.8	6.4	6.9	7.4	7.8	8.3	8.7	9.0
100	3.6	4.4	5.0	5.6	6.2	6.7	7.1	7.6	8.0	8.4	8.7
	12-inch Diameter Conduit										
20	7.8	9.5	11	12							
30	7.3	9.0	10	12	13	14					
40	7.0	8.5	10	11	12	13	14	15	16		
50	6.7	8.1	9.4	11	12	13	13	14	15	16	16
60	6.4	7.8	9.0	10	11	12	13	14	14	15	16
70	6.1	7.5	8.7	10	11	12	12	13	14	14	15
80	5.9	7.3	8.4	9.4	10	11	12	13	13	14	15
90	5.7	7.0	8.1	9.1	10	11	12	12	13	14	14
100	5.6	6.8	7.9	8.8	9.7	10	11	12	12	13	14
	14-inch Diameter Conduct										
20	10	12	14	16							
30	9.6	12	14	15	17	18					
40	9.2	11	13	14	16	17	18	19	20		
50	8.8	11	12	14	15	17	18	19	20	21	21
60	8.4	10	12	13	15	16	17	18	19	20	21
70	8.2	10	12	13	14	15	16	17	18	19	20
80	8.0	10	11	12	14	15	16	17	18	19	19
90	7.7	9	11	12	13	14	15	16	17	18	19
100	7.5	9	10	12	13	14	15	16	17	17	18

TABLE 9.5. *(Continued)*

Length of Conduit, Feet	Total Head, Feet										
	4	6	8	10	12	14	16	18	20	22	24
16-inch Diameter Conduit											
20	14	17	19	21							
30	13	16	18	20	22	24					
40	12	15	18	20	22	23	25	26	28		
50	12	15	17	19	21	22	24	25	27	28	29
60	12	14	16	18	20	22	23	25	26	27	28
70	11	14	16	18	19	21	22	24	25	26	27
80	11	13	15	17	19	20	22	23	24	26	27
90	11	13	15	17	18	20	21	22	24	25	26
100	10	13	15	16	18	19	21	22	23	24	25
18-inch Diameter Conduit											
20	17	21	25	28							
30	17	20	24	27	29	31					
40	16	20	23	26	28	30	32	34	36		
50	16	19	22	25	27	29	31	33	35	37	38
60	15	19	22	24	26	28	30	32	34	36	37
70	15	18	21	23	26	28	30	31	33	35	36
80	14	18	20	23	25	27	29	30	32	34	35
90	14	17	20	22	24	26	28	30	31	33	34
100	14	17	19	22	24	26	27	29	31	32	33
20-inch Diameter Conduit											
20	22	27	31	34							
30	21	26	30	33	36	39					
40	20	25	29	32	35	38	41	43	45		
50	20	24	28	31	34	37	40	42	44	46	49
60	19	24	27	30	33	36	38	41	43	45	47
70	19	23	26	30	32	35	37	40	42	44	46
80	18	22	26	29	32	34	36	39	41	43	45
90	18	22	25	28	31	33	36	38	40	42	44
100	17	21	25	28	30	32	35	37	39	41	43
24-inch Diameter Conduit											
20	32	39	45	51							
30	31	38	44	50	54	59					
40	30	37	43	48	53	57	61	65	68		
50	30	36	42	47	52	56	60	63	67	70	73
60	29	36	41	46	51	55	59	62	65	68	72
70	28	35	40	45	49	53	57	60	64	67	70
80	28	34	40	44	48	52	56	59	62	65	68
90	27	34	39	43	47	51	55	58	61	64	67
100	27	33	38	42	46	50	54	57	60	63	66
26-inch Diameter Conduit											
20	36	44	51	57							
30	35	43	49	55	60	65					
40	34	42	48	54	59	64	68	72	76		
50	33	41	47	53	58	62	67	71	74	78	81
60	33	40	46	52	56	61	65	69	73	76	80
70	32	39	45	50	55	59	64	68	71	75	78
80	31	38	44	49	54	58	62	66	70	73	76
90	31	37	43	48	53	57	61	65	69	72	75
100	30	36	42	47	52	56	60	64	67	71	74

TABLE 9.5. *(Continued)*

Length of Conduit, Feet	Total Head, Feet										
	4	6	8	10	12	14	16	18	20	22	24
					30-inch Diameter Conduit						
20	51	62	72	80							
30	50	61	70	78	86	93					
40	49	60	69	77	84	91	97	103	109		
50	48	58	68	75	83	89	96	101	107	112	117
60	47	57	66	74	81	88	94	99	105	110	115
70	46	56	65	73	80	86	92	98	103	108	113
80	45	55	64	72	78	85	91	96	101	106	111
90	44	54	63	70	77	83	89	94	99	104	109
100	44	53	62	69	76	82	88	93	98	103	107
					36-inch Diameter Conduit						
20	75	91	105	117							
30	73	89	103	115	126	136					
40	72	88	102	114	124	134	144	152	161		
50	71	87	100	112	123	131	142	150	158	166	173
60	70	86	98	110	121	130	140	148	156	163	171
70	69	84	97	109	119	129	138	146	154	161	168
80	68	83	96	107	117	127	135	144	151	159	166
90	67	82	95	106	116	125	134	142	149	157	164
100	66	81	93	104	114	123	132	140	147	155	162

Trash racks should be placed on morning glory and drop inlets if there is the possibility of trash large enough to plug the structure being washed from the watershed.

Guard rails should be used around the structure if deemed necessary for the protection of livestock or children.

Capacity of a pipe spillway is affected by the total head, the type of inlet, the length and diameter of the conduit, and the type of material. The head that will cause water to flow in a pipe spillway is equal to the total head minus entrance losses and friction losses in the conduit. For most pipe spillways, entrance losses are relatively small compared to friction losses in the conduit. The difference in entrance losses between the various types of inlets are so small that for practical purposes it can be assumed that the capacities of pipe spillways are the same regardless of the type of entrance. Manning's formula can be used to determine the capacity. Table 9.4 gives the capacity of pipe spillways with corrugated metal conduits. These were computed using a Manning's n of 0.025. Table 9.5 gives the capacities of pipe spillways with smooth metal conduits. Capacities were computed using a Manning's n of 0.013.

The total head is the difference in elevation between the water surface at the inlet and the center of the outlet end of the conduit, or the water surface, if the outlet end is submerged.

STRAIGHT DROP SPILLWAY

The straight drop spillway is a weir structure. Flow passes through the weir, drops to an approximately level apron or stilling basin, and

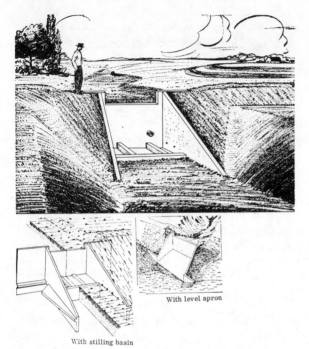

Figure 9.10. Straight drop spillway. (Courtesy U.S. Soil Conservation Service)

With level apron

With stilling basin

then passes into the downstream channel. It may be constructed of reinforced concrete, rock masonry, concrete blocks, steel or timber piling, or prefabricated metal (Fig. 9.10).

Adaptability. The straight drop spillway is adapted to the control of overfalls up to 8 feet where large rates of flow must be handled and an appreciable amount of detention storage cannot be provided. It may be used for the following purposes: (1) for grade stabilization in lower reaches of grassed waterways (Fig. 9.10); (2) for grade control in stabilizing channels; (3) as outlets for tile and surface water into drainage ditches; (4) for protection of the outlet end of a grassed waterway or sod chute; and (5) as an island type structure (Fig. 9.11) where low flows are carried by the drop spillway and high flows are diverted around the structure. This structure can be used only where there is sufficient area of nearly level land on either side of the structure to carry the overflow without damage to the land or the crop. If incoming water is loaded with sediment, the sediment may be deposited in the overflow area. This method has been used to build up a layer of silt loam sediment over heavy gumbo to improve its productivity.

Advantages and Limitations. The advantages of the straight drop spillway are that it is relatively stable if properly designed and installed, and it is not likely to be clogged by debris.

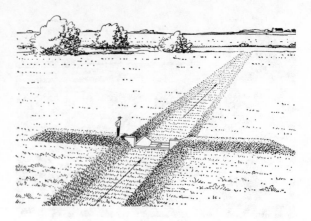

Figure 9.11. Island-type structure. (Courtesy U.S. Soil Conservation Service)

Its limitations include the following: (1) a stable grade below the structure is essential; (2) it is relatively expensive if the discharge capacity is less than 100 cubic feet per second (cfs) and the total head is over 8 feet; and (3) it is not a suitable structure if detention storage is needed.

Design and Construction. The capacity of drop spillways can be computed by the general weir formula

$$Q = CLh^{3/2}$$

The value of C varies with the type of spillway and the entrance conditions. For exact values of this coefficient see U.S. Soil Conservation Service (5). L is the length of the weir and h is the depth of flow over the weir. Table 9.6 can be used to obtain the approximate capacity of straight drop spillways if more specific information is not available.

Accepted construction practices should be followed in mixing and placing the concrete, selecting and placing the reinforcing, and curing the concrete.

CULVERT BOX INLET

The culvert box inlet is formed by placing a box inlet drop spillway at the upstream end of a culvert. It may be built as an integral part of a new culvert, or it may be fastened to the upstream headwall of an existing culvert (Fig. 9.12).

Adaptability. The culvert box inlet (1) controls gradients above culverts in either a natural or constructed channel; (2) controls gradients and reduces erosion in the ditches bordering the road or highway; (3) serves as an outlet for tile drains; and (4) may serve as a cattle ramp, with some modification in the design of the box, when the culvert is used as a cattle pass.

TABLE 9.6. Capacities of Straight Drop Spillways, cfs

Depth of Weir in Feet	Length of Weir in Feet										
	4	6	8	10	12	14	16	18	20	22	24
1.0	11	17	22	28	33	39	44	50	55	61	66
1.5	20	30	40	50	60	70	81	91	101	111	121
2.0	31	46	62	77	92	108	123	138	154	169	185
2.5		64	85	107	128	149	170	192	213	234	256
3.0		83	111	139	167	194	222	250	278	306	333
3.5			130	174	208	243	278	312	347	382	416
4.0			168	210	252	294	336	378	420	462	504

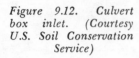

Figure 9.12. Culvert box inlet. (Courtesy U.S. Soil Conservation Service)

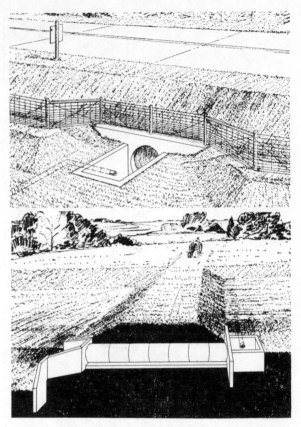

Advantages and Limitations. The culvert box inlet is one of the most economical structures for controlling overfalls because the existing culvert and highway embankment replace the outlet portion of the typical drop spillway. It does however require the availability of a structurally sound road culvert. It is often attached to a road culvert which is the property of a roadway governing body and in this instance requires their permission.

Design and Construction. The approximate capacity of culvert box inlets can be obtained from Table 9.7. A box inlet capacity need not be greater than 1.25 times the culvert capacity unless this is insufficient to meet the following requirements: to make the horizontal area of the box inlet at least 1.5 times the cross-sectional area of the culvert, and to make the dimensions of the box inlet large enough to prevent submergence of the existing culvert headwall at the design runoff or to extend the existing headwall.

At least one foot of freeboard should be allowed between the design

TABLE 9.7. Capacities of Box Inlets, cfs

Depth of Flow, Feet	Length of the Box Inlet, feet*													
	4	6	8	10	12	14	16	18	20	22	24	26	28	30
1.0	11	19	25	31	37	43	50	56	62	68	74	81	87	93
1.5	20	34	46	57	68	80	91	103	114	125	137	148	159	171
2.0	31	53	70	88	105	123	140	159	175	193	210	228	245	263
2.5		74	98	123	147	172	196	221	245	270	294	319	343	368
3.0		97	129	161	193	226	258	290	322	354	387	419	451	483
3.5		122	162	203	244	284	325	365	406	447	487	528	568	610
4.0		149	198	248	298	347	397	446	496	546	595	645	694	744

* The length of the wall over which water flows in entering the inlet.

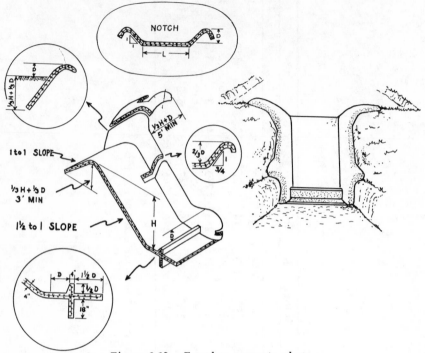

Figure 9.13. Formless concrete chute.

water level and the lowest point on the roadway, or the requirements of the highway department should be met.

If the culvert is of concrete, the box inlet may be constructed of reinforced concrete, plain concrete, or concrete blocks. Corrugated metal box inlets are commonly used if the culvert is of corrugated metal.

FORMLESS CONCRETE CHUTE

The formless concrete chute is a spillway constructed without the use of special forming material. The earth is excavated to the dimensions and shape of the structure. Concrete is placed on this earth form to the depth required and floated into shape. One type of formless concrete chute is shown in Figure 9.13.

Adaptability. This structure is best adapted to overfalls of less than 6 feet in height. It may be used to control overfalls in natural and constructed channels, to prevent erosion at the ends of waterways and diversions, and to lower runoff water over drainage ditch banks.

Advantages and Limitations. The use of earth as a form eliminates the need for purchasing forming materials; hand labor is required to form the earth to the desired shape and to place the concrete. This structure is limited to sites that have good natural drainage, and it should not be used as a water-impounding structure.

Design and Construction. The principal dimensions needed for construction are given in Figure 9.13. The size of the notch can be determined from Table 9.8. If the width of notch is greater than 8 feet, or if the height of the overfall is over 6 feet, the thickness of the concrete in the floor of the structure is increased to 5 inches and additional reinforcing is added.

The earth form should be smooth and shaped to the proper grade to prevent the use of an excessive amount of concrete. The soil must be damp and firm to provide a good base for the concrete. Concrete with a twenty-eight-day strength of at least 2,500 pounds per square inch is used. A stiff mix is needed to prevent the concrete from slumping down the slopes. Precautions should be taken during construction to assure that the proper thickness of concrete is obtained, with the reinforcing mesh placed as indicated in Figure 9.13. The concrete is worked into proper shape with a wooden float. A curing compound should be used or the structure covered with straw, plastic, or other material and kept moist for at least one week following construction.

EMERGENCY SPILLWAY

Emergency spillways are provided to convey the design flow around a structure at a nonerosive velocity to a safe point of release. They may be used in conjunction with ponds, irrigation reservoirs, and stabilization structures, with earth embankment structures that provide an appreciable amount of detention storage, or with full-flow structures that provide little detention storage.

For some structures with favorable spillway conditions the emergency spillway may carry all the runoff. For structures with less favorable spillway conditions, a principal spillway can be used to carry the low rates of runoff and an emergency spillway provided to carry the balance. The emergency spillway should be placed above the principal spillway a distance equal to or greater than the minimum entrance head required to cause the principal spillway to flow full.

EARTH EMBANKMENT STRUCTURES

Emergency spillways for this type of structure usually consist of an approach channel, control section, and exit section as shown in Figure 9.14.

If a pipe spillway is located in conjunction with an emergency spillway, alternative locations for the pipe should be considered as indicated in Figure 9.15. If located at *A,* the pipe is placed in the earth fill. A shorter pipe is usually required; however, extra precaution must be taken to obtain good compaction around the pipe and to provide necessary

TABLE 9.8. Capacities of Formless Concrete Chutes

Capacities in Cubic Feet Per Second

Length of Notch – Feet – Measured Thus:

Computed from Formula
$$Q = 3.75LD^{1.55} + 3.2D^{2.6}$$

Depth of Notch feet	2	3	4	5	6	7	8	9	10	12	14	16	18	20	22	24
1.0	11	15	18	22	26	30	33	37	41	48	56	63	71	78	86	93
1.25	16	22	27	32	38	43	48	53	59	69	80	91	101	112	122	133
1.50	23	30	37	44	51	58	65	73	80	94	108	122	136	150	164	178
1.75	32	41	49	58	67	76	85	94	103	121	139	157	174	192	210	228
2.00	41	52	63	74	85	96	107	118	129	151	173	195	217	239	261	283
2.25	53	66	79	92	105	119	132	145	158	185	211	237	264	290	316	343
2.50	66	81	97	112	128	143	159	174	190	221	252	283	314	345	376	407
2.75	80	98	116	134	152	170	188	206	224	260	296	332	368	404	440	476
3.00	97	117	138	159	179	200	220	241	262	303	344	385	426	467	509	550

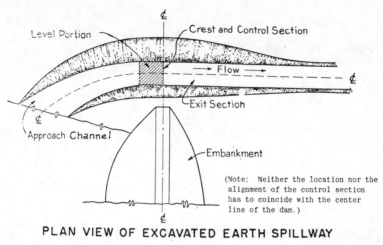

PLAN VIEW OF EXCAVATED EARTH SPILLWAY

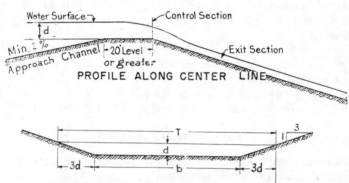

PROFILE ALONG CENTER LINE

CROSS SECTION AT CONTROL SECTION

Figure 9.14. Emergency spillway used in conjunction with earth embankment structures. (Courtesy U.S. Soil Conservation Service)

antiseep collars to prevent seepage along it which could result in washout of the pipe and embankment. If located at *B,* the pipe is placed behind the embankment, and a longer pipe is usually required. However, less damage would result from seepage along the pipe.

At locations where the end of the embankment is near the top of a ridge, the runoff may be discharged across the ridge onto the opposite slope as shown in Figure 9.16 if legal problems will not be involved. In this case, the 20-foot level section is not required. More often the runoff from the control section must be discharged onto sloping ground below the structure, in which case an exit section consisting of a grassed waterway or a diversion channel should be provided. If the land slope below the control section is not too steep, the runoff from the control section may be discharged into a grassed waterway of approximately the

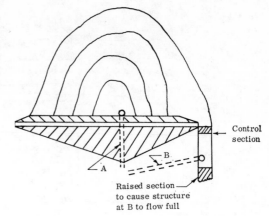

Figure 9.15. Alternative locations for principal spillways used in earth embankment structures.

Control
section

Raised section
to cause structure
at B to flow full

same width as the control section (Fig. 9.17). If the land slope below the control section is quite steep, the waterway will be much wider than the control section, and a layout similar to the one in Figure 9.18 should be used.

The runoff from the structure may be routed into a diversion channel and carried to a location at some distance from the embankment to be discharged (Fig. 9.19). A diversion may also discharge into a grassed waterway or a structure. The 20-foot level section is not required if the diversion starts at the end of the embankment.

FULL-FLOW STRUCTURES

There are several possible locations for emergency spillways used in conjunction with full-flow structures.

In Figure 9.20 runoff from a grassed waterway discharges into a diversion which carries the runoff to the structure and the emergency spillway. When the water reaches a certain level in the diversion channel, it flows over the crest of the emergency spillway onto the natural slope. The diversion in this spillway section is designed so that the

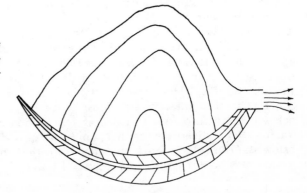

Figure 9.16. Runoff discharges across ridge into another drainageway.

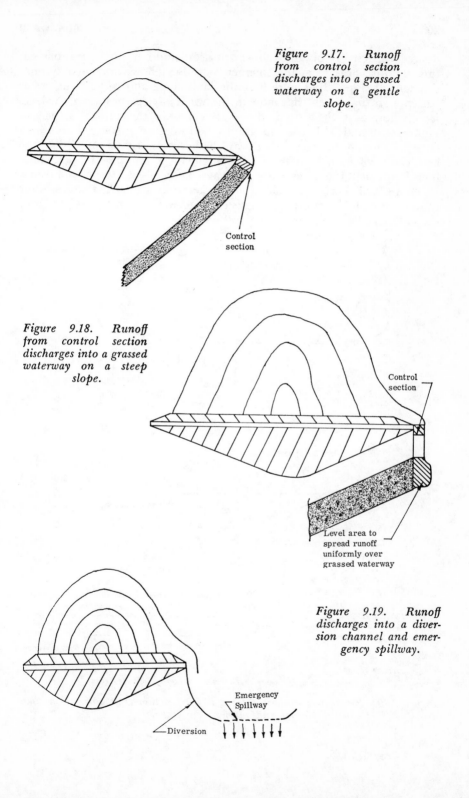

Figure 9.17. Runoff from control section discharges into a grassed waterway on a gentle slope.

Control section

Figure 9.18. Runoff from control section discharges into a grassed waterway on a steep slope.

Control section

Level area to spread runoff uniformly over grassed waterway

Figure 9.19. Runoff discharges into a diversion channel and emergency spillway.

Emergency Spillway

Diversion

desired capacity is obtained by excavation and no ridge is provided; thus the discharge from the emergency spillway flows immediately onto the natural slope. In Figure 9.21 the land slope above the structure is flat enough that when the water in the waterway reaches a certain level it will flow over the side of the waterway onto the natural slope. In Figures 9.20 and 9.21, the topography and vegetative cover in the area onto which the emergency spillway discharges must be such that the flow can be handled safely. Good judgment is required to determine how frequently the emergency spillway can be used at such locations. In figure 9.22 the natural slope and vegetation are such that the flow cannot be handled safely, so a grassed waterway is constructed to carry the discharge from the emergency spillway.

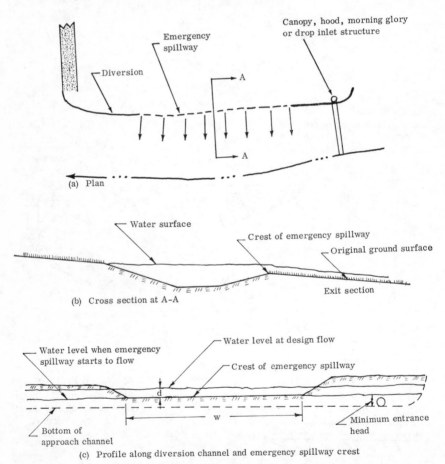

Figure 9.20. *Emergency spillway located in conjunction with a diversion chan-nel and full-flow structure.*

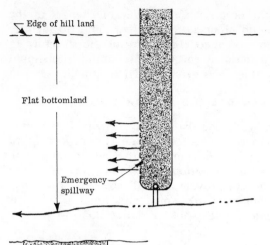

Figure 9.21. Possible location for an emergency spillway if the slope in the grassed waterway is flat enough that the crest of the emergency spillway can be constructed level throughout its length.

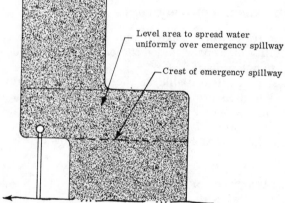

Figure 9.22. Use of a grassed waterway to carry the runoff from an emergency spillway for a full-flow structure.

EARTH EMBANKMENTS

In designing and constructing earth embankments for farm ponds, irrigation reservoirs, and grade stabilization structures, several factors should be considered.

SITE INVESTIGATION

The suitability of a pond or reservoir site is dependent on the ability of the soils in the reservoir area to hold water and to provide a stable foundation for the embankment. In areas where the nature of the soil is such that permeable or unstable materials may be present in the soil profile, a geologic investigation should be made to determine if a satisfactory structure can be built at that site. Soil borings or test pits should be made at intervals over the structure site. The frequency of the borings depends on the occurrence of significant changes in the soil profile. The borings should be made to sufficient depth to identify material that may be unstable or affect the ability of the structure to hold water. If undesirable material is encountered, another site should be considered or the undesirable areas given special treatment.

PREPARATION OF SITE AND BORROW AREA

All trees, stumps, trash, brush, sod, large roots, perishable material, and loose soil should be removed from these areas. Topsoil is stripped from the embankment foundation and borrow areas and deposited in storage piles. After the stripping operation, the ground surface within the foundation area is scarified to provide a bond between the foundation and earth fill. Overhanging banks, pits, and holes within the foundation area are graded to a one-to-one or flatter slope to provide a bond with the fill.

CORE TRENCH CONSTRUCTION

The core trench is excavated to the extent and dimensions necessary to prevent any possibility of seepage under the embankment. Side slopes are constructed one to one or flatter.

The most impervious material available at the site is utilized to backfill the core trench. The moisture content of the backfill material should be sufficient to secure proper compaction. (When kneaded in the hand, the soil should just form a ball which does not readily separate.) The trench must be kept free of standing water during backfill operations.

INSTALLATION OF BLANKET

Pond basins containing a high percentage of coarse-grained soils may require a blanket of less permeable material to prevent seepage. At some locations the topsoil may be less permeable than the subsoil. The less permeable material in the area should be stockpiled and spread as a blanket over the more permeable material in the pond basin. The blanket should be placed in 4-inch, well-compacted layers to a minimum thickness of 12 inches in most cases. If a blanket is used over the entire pond basin, a core trench is not required.

INSTALLATION OF PRINCIPAL SPILLWAY CONDUIT
AND STOCK WATER PIPE

The conduit for the principal spillway and the pipe for the stock-watering system, if needed, are placed on solid foundations either in trenches excavated in undisturbed soil or on fill material placed in 6-inch layers and properly compacted. Antiseep collars are installed on the conduit and on the stock water pipe if needed. If no cradle is used under the conduit, the foundation is shaped to fit the conduit for a depth equal to one-tenth of its diameter. A uniformly firm bed is necessary throughout its length.

DESIGN

Top Width. The following minimum top widths are suggested for dams of various heights: with a dam height of under 10 feet, the minimum top width should be 8 feet; with a height of 10–20 feet, the minimum top width should be 10 feet; and with a height of 21–30 feet, the minimum top width should be 12 feet. If the top of the embankment is to be used as a roadway, the top width should be at least 12 feet.

Side Slopes. The side slopes on the settled embankment should not be steeper than three to one on the upstream side and two to one on the downstream side, unless based on previous favorable results on similar soils or on soil mechanics tests for stability. Flatter slopes should be used if necessary to insure the stability of the embankment. On the downstream side, flatter slopes are also desirable if mechanical equipment is to be used for mowing and maintenance.

Slope Protection. When the surface area of the reservoir exceeds 5 acres, extra precaution should be taken to reduce damage caused by wave action. There are several methods that may be considered: an 8-foot berm can be constructed on the upstream face of the embankment 0.5 feet above the crest of the principal spillway; a portion of the upstream slope of the embankment can be riprapped; or the upstream slope of the embankment can be constructed to a four-to-one slope. Reed canary or other suitable grass or aquatic plants is used to protect berms from wave action. The upstream slopes of reservoirs smaller than 5 acres are protected from wave action by reed canary or other suitable grass, aquatic plants, riprap, or other means as site conditions indicate.

Settlement. Sufficient overfill should be provided during construction to allow for settlement. If the material is placed in 6-inch layers at a moisture content sufficient for good compaction, 5 percent is added if compacted thoroughly with wheel scrapers or with rollers; 10 percent if a bulldozer only is used. For material which is quite dry or excessively wet, suggestions are given below under Construction.

Freeboard. The top of the embankment should be constructed high enough to prevent waves or runoff from storms greater than the design frequency from overtopping the embankment. Freeboard is the difference in elevation between the water level in the reservoir when the spillway is flowing at the designed depth and the top of the settled fill. If both principal and emergency spillways are used and the reservoir is less than

660 feet long, a minimum freeboard of 1.0 foot should be provided. For reservoirs between 660 and 1,320 feet long, the minimum freeboard should be 1.5 feet, and for reservoirs over 1,320 feet long, it should be 2.0 feet (5). If only an emergency spillway is used, add 0.5 foot to the minimum freeboards suggested above.

STAKING OUT

A tentative location for the embankment is selected and a row of stakes is set marking the proposed location of the center line. The shape required to store the greatest volume of water for the amount of earth moved will vary from a straight embankment constructed across a relatively deep depression to one semicircular in shape constructed on a relatively uniform slope.

A tentative location for the waterline is selected and a series of stakes set at this elevation. The area enclosed by these stakes and by a line marking the intersection of the water level and the embankment is determined. If this surface area does not equal that desired, the water level stakes may be raised or lowered or the location of the embankment may be shifted to give the size pool desired.

The design runoff to be expected from the drainage area is computed. The spillway size and the depth of water expected during the design runoff is determined. The desired freeboard is added and the elevation of the top of the embankment computed.

Slope stakes outlining the base of the embankment are set according to procedures in Chapter 14.

VOLUME OF EARTH

Methods for computing the volume of earth in embankments are discussed in Chapter 13.

CONSTRUCTION

Fill material must not contain any appreciable concentration of vegetation, roots, large rocks, frozen soil, or other foreign substances. Moisture content should be sufficient to secure proper compaction. (When kneaded in the hand, the soil should just form a ball which does not readily separate.)

The fill material is placed in approximately 6-inch layers that extend over the full width and length of the dam. Each layer should be compacted by tractors, earth moving equipment, or rollers. If the moisture content is deficient, water is sprinkled on the surface of each lift of material and thoroughly mixed and compacted prior to placing additional fill. If the moisture content of the material is excessive, it should not be used until it has air dried to proper moisture content. Sufficient overfill is used to allow for settlement.

Backfill adjacent to the pipe spillway and stock-watering pipe must be free of rocks, clods, or clumps. It is placed in 4-inch layers and tamped by hand or with mechanical tampers. Rubber-tired tractors may be used to supplement the tamping along the sides of the pipe and to provide a berm of compacted earth equal to the pipe diameter on each side. Backfill is brought up approximately equal on each side of the pipe to prevent side movement of the pipe. Care should be taken to prevent the pipe from uplifting when backfilling under the haunches. Heavy equipment should not be driven over the pipe until a minimum of 2 feet of compacted fill has been placed on top of it.

Frozen material should not be used in a fill, and fill material is not to be placed on frozen earth. When fill material starts to freeze during placement, construction is stopped until proper temperatures prevail. When it is desired to place fill on an area that is frozen, the frozen soil is permitted to thaw completely and to dry to the proper moisture content before placement, or the frozen material is completely removed before placing additional fill.

At sites where there is a limited amount of the most impervious fill material, this material is placed in the core trench and the center section of the embankment, and the more permeable material is placed in the downstream part of the embankment.

The topsoil saved during the site preparation is placed as a top dressing on the surface of the earth fill and emergency spillway.

VEGETATION

Exposed surfaces of the earth spillway, embankment, and borrow area not covered by the permanent pool should be fertilized, seeded, and mulched to provide a good vegetative cover.

SEEPAGE

Figure 9.23 illustrates situations in which seepage may occur from a water storage structure.

Figure 9.24 indicates precautions which should be taken during site selection, investigation, and construction to prevent seepage losses. Following is a listing of possibilities for seepage indicated by numbers encircled in Figure 9.23 and the suggested precautions or solutions:

1. Seepage through strata or pockets of permeable material. These may or may not be exposed during excavation. *Solution:* Make a thorough site investigation prior to construction. Either move to another site or give special treatment to undesirable areas such as the compacted blanket of impervious material in Figure 9.24.
2. Seepage along roots and root cavities. If trees in the vicinity of the embankment are cut, the roots will die, leaving cavities along which water will flow. *Solution:* Remove all roots from the embankment

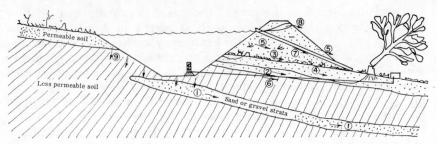

Figure 9.23. Situations in which seepage may occur from a water storage structure.

area prior to construction.

3. Seepage along the plane between the original ground and the embankment. *Solution:* Follow suggestions given under Preparation of Site and Borrow Area.

4. Seepage under the embankment and through a layer of permeable material. *Solution:* Block flow by construction of a core trench as suggested under Core Trench Construction.

5. Seepage through the embankment. *Solution:* Build the embankment with proper top width and side slopes; remove all brush, roots, and debris from the borrow area, so it will not be deposited in the embankment; place the less permeable material on the water side of the embankment and more permeable material in the downstream part; place fill in thin layers and compact thoroughly; and follow other suggestions given under Construction.

6. Seepage along stock water pipe. *Solution:* Install antiseep collars and compact earth around pipe.

7. Seepage along principal spillway conduit. *Solution:* Install antiseep collars and follow suggestions given under Installation of Principal Spillway Conduit and Stock Water Pipe.

8. Flow through muskrat burrows and cavities created by other burrowing animals. *Solution:* Build embankment with proper top width and side slopes; design the spillways to reduce fluctuation of water

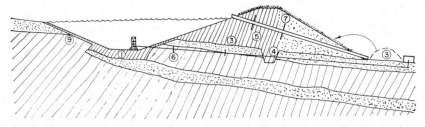

Figure 9.24. Precautions to take during construction of a water storage structure to prevent seepage.

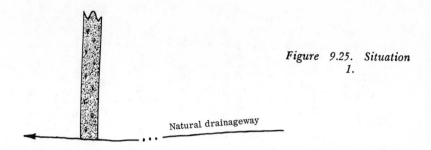

Figure 9.25. Situation
1.

Natural drainageway

level to a minimum; and keep the embankment cleared of brush and debris.

9. At sites where the soil is permeable throughout the profile there may be seepage over the entire basin. *Solution:* Scarify the basin area to a depth of 8–10 inches; compact the loosened soil at optimum moisture content to a dense layer; and on more permeable soils install a blanket as discussed previously.

SELECTING THE PROPER STRUCTURE

In any given situation there are usually alternative methods of control and different types of structures that could be used. A number of different situations will be presented and alternative solutions discussed.

Situation 1: A single waterway, diversion, or terrace discharges into a natural drainageway (Fig. 9.25).

Solution: A number of different conditions and the structures that best fit the need in each case are presented in Table 9.9.

Situation 2: Several waterways discharge into a natural drainageway (Fig. 9.26).

Solution 1: Place a structure at the end of each waterway.

Solution 2: Connect two or more waterways with a diversion and carry water to a common structure.

Solution 3: Place an embankment across the natural drainageway. Install principal and emergency spillways to carry the runoff from the waterways plus the natural drainageway. This solution will provide a water supply, a crossing over the main drainageway, and some stabilization for the natural drainageway, in addition to stabilizing the waterways. However, it may become quite expensive if the natural drainageway is carrying runoff from a large area. If a crossing over the main drainageway or a water supply is not needed,

TABLE 9.9. Structure Selection Chart

	Natural Drainageway		Over-Fall,* Feet	Structure Discharge cfs	Possibilities for Emergency Spillway	Purpose of Structure	Other Considerations	Structure Choices
	Condition	Size, Acres						
Case #1	Stable	50–200	Under 4	Under 50	Good	Stabilization of outlet	Deep fertile soil flat slope in outlet	Slope to natural drain and establish sod
Case #2	Stable	50–200	4–8	Under 50	Good	Stabilization of outlet	Shallow soil, steep slope in outlet	Straight drop spillway, formless flume, or pipe spillway**
Case #3	Unstable	50–200	4–8	0–200	Good	Stabilization of outlet		Pipe spillway**
Case #4	Stable	50–200	4–8	Over 50	Poor	Stabilization of outlet		Straight drop spillway, formless concrete flume.
Case #5	Unstable	50–200	4–8	Over 50	Poor	Stabilization of outlet & natural drain, water supply, crossing		Earth embankment*** across main drain
Case #6	Stable	Over 200	4–8	Over 50	Poor	Stabilization of outlet		Straight drop spillway, formless concrete flume
Case #7	Unstable	Over 200	4–30	50–200	Good	Stabilization of outlet		Pipe spillway**

*The overfall is the difference in elevation between the emergency spillway and the bottom of the natural drainageway immediately below the structure.

**Pipe spillway may have a canopy, hood, morning glory or drop inlet plus an emergency spillway where feasible.

***Earth Embankment Structure will consist of the earth embankment; a pipe spillway, which may have a canopy, hood, morning glory or drop inlet; and an emergency spillway.

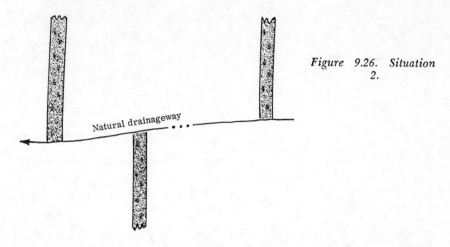

Figure 9.26. Situation 2.

the solution should be selected which will give the desired stabilization at the minimum cost.

Situation 3: Runoff from hill land floods across the bottomland. Runoff from 5 acres is concentrated at *A*, from 25 acres at *B*, and from 5 acres at *C* (Fig. 9.27).

Solution: Intercept with a diversion at the edge of the hill land. Discharge diversion into a grassed waterway or drainage ditch across bottomland. Use a pipe spillway structure to discharge water into the river.

Situation 4: Runoff from 10 acres is concentrated at *A*, from 100 acres at *B*, and from 15 acres at *C* (Fig. 9.27).

Solution: Place an earth embankment with a pipe spillway and emergency spillway at *B* to collect sediment and reduce the runoff to a rate that can be carried across the bottom without causing flood

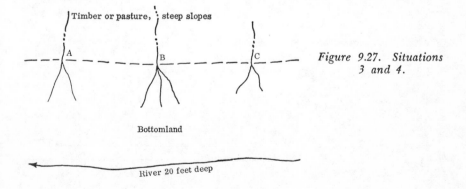

Figure 9.27. Situations 3 and 4.

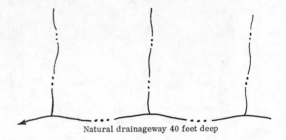

Figure 9.28. Situation 5.

Natural drainageway 40 feet deep

damage. Intercept runoff at *A* and *C* with diversions, and discharge into structure at *B*. Use a grassed waterway or drainage ditch to carry discharge from the structure across bottomland. Use a pipe spillway structure to discharge water into the river.

Situation 5: Gullies are cutting back from a deep natural drainageway by overfall erosion (Fig. 9.28).

Solution 1: Terrace land above the gully heads and divert water to a stable outlet. If not feasible to terrace, use a diversion to intercept the runoff from the gully heads and divert to a stable outlet.

Solution 2: Place a pipe spillway at the head of the gully that is most stable and provides the best structure site. Use diversions to intercept runoff from the other gullies and divert into the structures.

Solution 3: Place a pipe spillway structure at the head of each gully.

Situation 6: Active gullies are advancing through the field and along the road ditches (Fig. 9.29).

Solution: Terrace the area and eliminate the gullies in the field. Discharge terraces into a grassed waterway. Use a culvert box inlet on the end of the road culvert to stabilize the lower end of the waterway and to reduce erosion in the road ditches.

Figure 9.29. Situation 6.

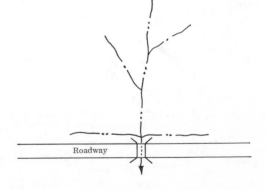

Roadway

SPILLWAY DESIGN

The spillway for a structure may consist of a principal spillway, an emergency spillway, or a combination of both. The combined spillway capacity should be great enough to carry the runoff from the design storm without causing excessive erosion.

THE DESIGN STORM

The following factors should be considered in selecting the design storm: the additional cost of the spillways if designed to carry a larger rate of runoff, the probable extent of damage to the structure and to other property if the capacity were exceeded, the inconvenience of making repairs, and additional damage that might occur while the structure is being repaired or replaced.

Normally a fifty-year frequency runoff from the watershed should be used to determine the capacity of the spillways if any of the following conditions exist:

1. The height of the embankment, measured from the lowest point along the proposed center line to the top of the fill, is over 25 feet, and the fifty-year peak rate of runoff from the watershed is greater than 250 cfs.
2. The height of the embankment is less than 25 feet and the fifty-year peak rate of runoff from the watershed is greater than 600 cfs.
3. The product of the reservoir depth (feet) times the storage capacity below the crest of the emergency spillway (acre-feet) is less than 3,000. The reservoir depth is the difference in elevation between the emergency spillway crest and the lowest point in the original cross section along the center line of the dam.
4. The site is located above farmstead, homes, or other improvements where failure may cause extensive damage.

At least a twenty-five-year frequency runoff from the watershed should be used to determine the capacity of the spillways if any of the following conditions exist:

1. The height of the embankment is 25 feet or less and the fifty-year peak rate of runoff from the watershed is less than 600 cfs.
2. The height of the embankment is greater than 25 feet and the fifty-year peak rate of runoff from the watershed is less than 250 cfs.
3. The product of reservoir depth (feet) times the storage capacity (acre-feet) is less than 1,500.
4. The total height of the overfall in full-flow structures is greater than 25 feet. The overfall is the difference in elevation between the emergency spillway and the bottom of the natural drainageway immediately below the structure.

At least a ten-year frequency runoff from the watershed should be used to determine the capacity of the spillways if any of the following conditions exist:

1. The ten-year peak rate of runoff from the watershed is 80 cfs or less.
2. The product of reservoir depth (feet) times the storage capacity (acre-feet) is less than 500.
3. The height of the embankment or the total overfall in full-flow structures is 25 feet or less.

FREQUENCY OF USE

The frequency at which runoff may safely be passed through the emergency spillway is the primary factor influencing the design of the principal and emergency spillways. Good judgment determines how frequently the emergency spillway should be used. The maximum permissible velocity that the vegetation and soil on the area over which the runoff discharges can withstand should be the principal controlling factor. For example, at one location it may be possible to carry all the runoff through the emergency at less than the permissible velocity if a wide grassed waterway is constructed and an excellent sod maintained. A safer and less expensive solution may be to install a principal spillway and use the emergency less frequently. The merits, feasibility, and expense of alternatives should be considered at each location.

In addition to the velocity in the spillway, the condition of the topography and the vegetation of the area onto which the runoff discharges should be considered. For example, in Figure 9.20 the steepness of the bank of the stream into which the runoff will eventually discharge, the type of cover on the bank, and the width of the area over which the runoff will enter the stream must be considered in determining the frequency of use of the emergency spillway.

For some structures with favorable spillway conditions, the emergency spillway may carry all the runoff. The topography at these sites must be such that there is a smooth slope to a stable outlet, and the soil must be such that a good dense sod can be maintained. Factors such as an overly large watershed, springs or seeps, or wave action caused by size and location of the reservoir which will result in prolonged flow following a rainfall should be avoided. The velocity in the spillway should not exceed those given in Table 9.10(a).

At sites where it is not feasible to keep the velocities in the emergency spillway below those specified in Table 9.10(a) or where there is the possibility of prolonged runoff through the spillway following a rain, a principal spillway should be provided. Any flow in excess of that carried by the principal spillway or stored in detention storage will pass through the emergency spillway. If a principal spillway plus detention storage were designed to carry the maximum rate of runoff

TABLE 9.10. Maximum Permissible Velocities in the Exit Section of Emergency Spillways, Feet per Second

Quality of Vegetation	Less Erodible Soils[a] Channel slope (%)		More Erodible Soils[b] Channel slope (%)	
	0–7	8–16[c]	0–7	8–16[c]
(a) All runoff through emergency spillway				
Fair	4	3	3	[d]
Good	5	4	4	3
Excellent	6	5	5	4

	0–8	9–14	15–20	0–8	9–14	15–20
(b) Emergency used once a year or more frequently						
Fair	6	5	4	5	4	3
Good	7	6	5	6	5	4
Excellent	8	7	6	7	6	5
(c) Emergency used once each two years or less frequently						
Fair	7	6	5	6	5	4
Good	8	7	6	7	6	5
Excellent	10	8	7	8	7	6

[a] Less erodible soils are generally those with a higher clay content and higher plasticity. Typical soil textures are silty clay, sandy clay, and clay.

[b] More erodible soils are generally those that have a high content of fine sand or silt and lower plasticity. Typical soil textures are fine sand, silt, sandy loam, and silty loam.

[c] For slopes over 16 percent, use a principal spillway.

[d] Use a principal spillway.

to be expected in a one-half-year period, runoff would be expected to flow through the emergency spillway an average of twice a year.

If the emergency spillway is to be used one or more times a year, the velocity in the spillway should not exceed the velocities given in Table 9.10(b). If the design flow is in excess of 100 cfs, the spillway should have a gradual slope to a stable outlet.

If the emergency spillway is to be used once each two years or less frequently, the maximum velocity in the spillway should not exceed those in table 9.10(c). If the design flow is in excess of 150 cfs, the spillway should have a gradual slope to a stable outlet.

Occasionally it is not feasible to discharge the flow from the emergency spillway onto a stable outlet. For example, it may be necessary to discharge the flow over the bank of a natural drainageway. Good judgment based on experience is valuable in selecting the desired frequency.

If the bank has a stable slope and good vegetative cover, if the water from the emergency spillway tends to spread over a greater width than the spillway before flowing over the bank, and if the area between the bank and the spillway has a good vegetative cover, the following frequencies are suggested: the emergency spillway can be used once each six months if the design flow is less than 50 cfs and the design depth of flow over the crest of the spillway is less than 0.4 foot. It may be used once each year if the design flow is less than 100 cfs and the design depth of flow over the crest of the spillway is less than 0.6 foot. It may be used once each two years if the design flow is less than 150 cfs and the design depth of flow over the crest of the spillway is less than 0.8 foot.

The emergency spillway can be used once each five years if the bank is stable and has minor irregularities in slope and fair vegetation, if the water tends to spread over a width somewhat greater than the spillway before flowing over the bank, and if the area between the bank and the spillway has a good vegetative cover.

The emergency spillway should only be used once each ten years if the bank is unstable, the vegetation on the bank is poor, the water cannot spread over a greater width than the spillway before flowing over the bank, and the area between the bank and the spillway has only a fair vegetative cover.

SIZE OF EMERGENCY SPILLWAY

The emergency spillway should have the capacity to carry the runoff resulting from the design storm reduced by the discharge carried by the principal spillway or by the volume retained in detention storage or both.

Earth Embankment Structures. Width and depth of the control section. The control section is constructed to a trapezoidal shape with three-to-one or flatter side slopes, unless constructed in rock. When it is necessary to cross the emergency spillway with machinery the side slopes should be six to one or flatter. Ten feet is the minimum width for all control sections. For structures with a drainage area of 10 acres or less, the ten-foot minimum width for the control section should be used. For drainage areas larger than 10 acres, the required capacity is computed and the minimum width of the control section and depth of flow are obtained from Table 9.11.

In most cases it is advisable to make the control section as wide as is feasible. The wider control section will result in a shallower depth of flow, a lower velocity, and less fluctuation in the water level. The height of the embankment will also be less as may be the cost of construction.

Width and depth of the exit section. If the exit section is being designed as a grassed waterway, velocities may be selected from Table 9.10, and the width and depth determined by the procedure given in Chapter 6.

If it is being designed as a diversion, the control section is usually eliminated. The diversion is started at the water level in the pool. Sufficient freeboard should be provided above the design flow level in the diversion.

TABLE 9.11. Discharge through Control Section of Vegetated Spillway, cfs

Depth of Flow, Feet	Bottom Width of Control Section, Feet															
	10	12	14	16	18	20	22	24	26	28	30	32	34	36	38	40
0.5	6	8	9	11	12	13	14	16	18	20	21	23	25	27	28	29
0.6	10	12	13	15	17	19	20	22	24	27	29	32	34	36	38	40
0.7	13	15	17	20	22	25	27	29	31	34	37	40	43	46	49	51
0.8	17	20	23	26	29	32	34	37	40	43	46	50	54	57	60	63
0.9	20	24	28	32	35	39	42	46	49	52	56	61	66	70	74	77
1.0	25	30	34	38	42	47	51	55	59	63	67	72	77	82	87	92
1.1	30	35	40	45	50	55	60	65	70	75	80	85	90	96	102	108
1.2	35	41	47	53	59	65	71	76	81	87	92	98	104	110	117	124
1.3	41	48	55	62	68	75	81	87	93	100	107	114	120	126	133	140
1.4	47	55	62	70	77	85	91	98	106	114	122	130	136	143	151	160
1.5	54	62	70	79	87	95	103	111	119	128	137	146	153	161	170	180
1.6	61	70	79	88	97	106	115	125	124	144	153	162	171	180	190	200
1.7	68	78	88	98	108	118	128	138	148	158	168	178	188	200	210	220
1.8	76	86	97	108	119	130	141	153	164	175	185	197	208	220	231	242
1.9	83	95	106	118	130	142	154	167	179	192	203	216	228	240	252	264
2.0	91	104	116	128	141	154	167	181	194	208	221	235	248	261	274	288

If the runoff from the control section discharges onto a natural slope as in Figures 9.16 and 9.19, the topography of the area must be such that the water will spread over an area wide enough to keep the velocities below the limits for the existing vegetation, and the area must meet the conditions discussed previously under Frequency of Use.

Full-Flow Structures. Little detention storage is provided in full-flow structures. The emergency spillway must have the capacity to carry the peak rate of runoff from the design storm minus that carried by the principal spillway.

Width of the spillway crest. After the capacity has been determined, the width of the crest and the depth of flow can be determined from Table 9.12. If the runoff from the crest discharges onto a natural slope as in Figures 9.20 and 9.21, the topography of the area must allow the water to spread over an area wide enough to keep the velocity below the limits for the existing vegetation. The area must meet the conditions previously discussed under Frequency of Use.

The spillway crest should always be made as wide as practical at a given location. The wider spillway will have a shallower depth of flow with less velocity, and less height of fill around the principal spillway inlet will be required.

Slope of the spillway crest. The slope of the crest of the emergency spillway should allow the water to flow over at a uniform depth throughout its length. In most cases the crest can be made level. However in cases where the elevation of the water surface in the channel approaching the emergency spillway is appreciably less than at the structure, the spillway should be constructed with a slight slope, the crest being highest at the end toward the structure.

FLOOD ROUTING

Watershed runoff that is temporarily stored above the permanent pool level in a reservoir is known as detention storage. The peak rate of flow from the spillways of the impounding structure will be less than the peak rate of runoff from the watershed because of the detention storage. In the design, the combination of detention storage and spillway size should be selected that will result in the most economical and satisfactory structure.

The abbreviated flood-routing procedure presented in this chapter is based on typical watershed and structure characteristics. This procedure may be used for design of structures on watersheds up to 200 acres in size where serious property damage would not result from failure. For other structures, accepted flood-routing procedures based on the actual watershed and structure characteristics are used (4).

TABLE 9.12. Discharge through Emergency Spillways for Full-Flow Structures, cfs

Depth of Flow, Feet	Width of Spillway Crest – Feet														
	10	20	30	40	50	60	70	80	90	100	120	140	160	180	200
0.3	4	9	14	18	23	27	32	36	41	45	54	63	72	81	90
0.4	7	14	21	28	35	42	49	56	63	70	83	97	111	125	139
0.5	10	19	29	39	49	58	68	78	88	97	117	136	156	175	194
0.6	13	26	38	51	64	77	89	102	115	128	153	179	205	230	256
0.7	16	32	48	64	81	97	113	129	145	161	193	225	258	290	322
0.8	20	39	59	79	98	118	138	157	177	198	236	275	315	354	394
0.9	23	47	70	94	117	141	164	188	211	235	282	329	376	423	470
1.0	28	55	83	110	138	165	193	220	248	275	330	385	440	495	
1.1	32	63	95	127	159	190	222	254	286	317	381	444			
1.2	36	72	108	145	181	217	253	289	325	361	434				
1.3	41	82	122	163	204	245	285	326	367	408	489				
1.4	46	91	137	182	228	273	319	364	410	456					
1.5	51	101	152	202	253	303	354	404	455						

Because of the large number of symbols used they are listed and defined below:

Q_{ip} = peak rate of runoff from the watershed used in designing the principal spillway, in cfs.

Q_i = peak rate of runoff from the watershed used in designing both spillways (principal and emergency), in cfs.

V_{rp} = volume of runoff from the watershed used in designing the principal spillway, in acre-feet.

V_r = volume of runoff from the watershed used in designing both spillways, in acre-feet.

V_{sp} = volume of detention storage used in designing the principal spillway, in acre-feet

V_s = total volume of detention storage used in designing both spillways, in acre-feet.

Q_{op} = flow through the principal spillway, in cfs.

Q_{oe} = flow through the emergency spillway, in cfs.

PRINCIPAL SPILLWAY DISCHARGE

The relationship between detention storage and principal spillway discharge is determined by the following procedure:

1. Determine the frequency of storm for which the principal spillway should be designed.
2. Compute the peak rate of runoff expected from the watershed for this frequency storm (Q_{ip}). (See Ch. 5.)
3. Determine the volume of runoff expected from the watershed (V_{rp}) for this frequency storm. Multiply the value from Table 9.13 by the number of acres in the watershed. For watersheds in locations other than along line 1.0 in Figure 5.9, multiply the value in Table 9.13 by the appropriate location factor from Figure 5.9. For that por-

TABLE 9.13. Volume of Runoff from a Watershed, Acre-Feet per Acre

$V^a \times I^b$	Frequency of Storm						
	½ year	1 year	2 year	5 year	10 year	25 year	50 year
0.5	.02	.03	.03	.05	.07	.10	.12
0.6	.02	.03	.04	.07	.09	.12	.14
0.7	.03	.04	.05	.08	.10	.14	.17
0.8	.04	.05	.06	.09	.12	.16	.19
0.9	.05	.06	.07	.11	.14	.18	.21
1.0	.05	.07	.08	.12	.16	.20	.23
1.1	.06	.08	.09	.14	.17	.22	.26
1.2	.07	.09	.10	.15	.19	.24	.28
1.3	.08	.10	.12	.17	.21	.27	.31

[a] The vegetative cover factor from Chapter 5.
[b] The soil infiltration factor from Chapter 5.

TABLE 9.14. Relationship of Q_{op}/Q_{ip} to V_{sp}/V_{rp} for Principal Spillway Design

V_{sp}/V_{rp}	Q_{op}/Q_{ip}									
	0.00	0.01	0.02	0.03	0.04	0.05	0.06	0.07	0.08	0.09
0.0	1.00	0.99	0.98	0.96	0.95	0.94	0.92	0.91	0.90	0.88
0.1	0.87	0.85	0.84	0.82	0.81	0.79	0.78	0.76	0.74	0.73
0.2	0.72	0.70	0.68	0.67	0.65	0.64	0.62	0.61	0.60	0.58
0.3	0.57	0.55	0.54	0.52	0.51	0.50	0.49	0.47	0.46	0.45
0.4	0.44	0.43	0.42	0.41	0.40	0.39	0.38	0.37	0.36	0.35
0.5	0.34	0.33	0.32	0.31	0.30	0.29	0.28	0.27	0.27	0.26
0.6	0.25	0.24	0.23	0.23	0.22	0.21	0.20	0.20	0.19	0.18
0.7	0.18	0.17	0.16	0.15	0.15	0.14	0.14	0.13	0.12	0.12
0.8	0.11	0.11	0.10	0.09	0.09	0.08	0.08	0.07	0.07	0.06
0.9	0.05	0.05	0.04	0.04	0.03	0.03	0.02	0.02	0.01	0.01

Source: USDA Soil Conservation Service, Engineering field manual for conservation practices, 1969.

tion of a watershed having slopes of 1 percent or less, the values from Table 9.13 are reduced by the following amounts: with a slope of 0–0.5 percent, reduce by 0.04 acre-feet per acre, and with a slope of 0.51–1.0 percent, reduce by 0.02 acre-feet per acre. If there are impoundments above the proposed structure, deduct the detention storage in these impoundments from the total runoff.

4. Determine the amount of detention storage (V_{sp}) between the water level at the point when flow starts through the principal spillway and the water level at the point when flow starts through the emergency spillway.
5. Compute V_{sp}/V_{rp}.
6. Using the computed value of V_{sp}/V_{rp}, determine the value of Q_{op}/Q_{ip} from Table 9.14.
7. Solve Q_{op}, the expected flow through the principal spillway.

Example: The following example is used to illustrate the effect of detention storage on the peak rate of flow from the spillways. Assume the following watershed and structure characteristics (the values given will be used in computing the peak rate and amount of runoff):

1. Drainage area—160 acres.
2. Location—northwest Missouri; $L = 1.04$.
3. Soil—Marshall silt loam; $I = 0.9$.
4. Average land slope on watershed—7 percent; $T = 0.96$.
5. Maximum distance runoff travels in reaching the structure—4,300 feet; $S = 1.0$.
6. Vegetation—120 acres of row crops planted on the contour with terraces and 40 acres of good quality meadow not terraced; $V = 0.88$, $C = 0.99$.
7. Terraces average 1,000 feet long; $P = 0.98$.
8. There is no appreciable surface storage on the watershed above the proposed structure.

9. The emergency spillway conditions are such that it can be used once each two years. A two-year frequency storm will be used in the design of the principal spillway; $F = 0.5$.

10. The site conditions indicate that a twenty-five-year frequency storm should be used in the design of the total structure; $F = 1.3$.

11. The reservoir pool will cover 2 acres when flow starts through the principal spillway.

12. The emergency spillway will be located at an elevation 2 feet above the level of the principal spillway.

13. The reservoir pool will cover 3 acres when flow starts through the emergency spillway.

14. The depth of flow through the emergency spillway should not exceed 1.8 feet.

15. The reservoir pool will cover 5.6 acres when the depth of flow in the emergency spillway is 1.8 feet.

16. The conduit for the principal spillway will be smooth iron pipe 80 feet long with a drop inlet. The total head causing flow will be 14 feet.

Computation: For determining the size of the principal spillway—

1. The principal spillway is designed for a two-year frequency storm.

2. The peak rate of runoff expected from the watershed for this two-year frequency storm is—

$$Q_{ip} = Q_t \times L \times I \times T \times S \times V \times C \times P \times F$$
$$Q_{ip} = 356 \times 1.04 \times 0.9 \times 0.96 \times 1.0 \times 0.88 \times 0.99 \times 0.98 \times 0.5$$
$$Q_{ip} = 137 \text{ cfs}$$

3. The volume of runoff expected from the two-year frequency storm is obtained from Table 9.13. The product of $V \times I$ (0.88×0.9) $= 0.79$. The volume of runoff in acre-feet per acre $= 0.06$ (Table 9.13). Correcting for watershed location (location factor is 1.04), $0.06 \times 1.04 = 0.062$ acre-feet per acre. The total volume of runoff $V_{rp} = 0.062 \times 160$ (acres in the watershed) $= 9.9$ acre-feet.

4. The amount of detention storage between the principal and emergency spillways is equal to the average pool area times the depth.

$$V = \frac{2 \text{ acres} + 3 \text{ acres}}{2} \times 2 = 5.0 \text{ acre-feet}$$

5. $V_{sp}/V_{rp} = 5.0/9.9 = 0.51$

6. $Q_{op}/Q_{ip} = 0.33$ (Table 9.14)

7. $Q_{op} = 0.33 \times 137 = 45$ cfs

8. The conduit for the principal spillway must have the capacity to carry at least 45 cfs. By referring to Table 9.5, we find that a 24-inch smooth pipe having a capacity of 52 cfs is required.

Other combinations of detention storage and pipe size could also be used. In designing a structure, the combination of pipe size and detention storage which gives the most economical and satisfactory structure should be selected.

EMERGENCY SPILLWAY DISCHARGE

The flow to the emergency spillway will be affected not only by the water in detention storage but also by the amount discharged through the principal spillway. The following procedure is used to determine the peak rate of flow through the emergency spillway:

1. Determine the frequency of storm for which the total structure should be designed.
2. Compute the peak rate of runoff expected from the watershed for this frequency storm (Q_i).
3. Determine the volume of runoff expected from the watershed (V_r). Multiply the value from Table 9.13 by the number of acres in the watershed. Make necessary corrections for watershed location and slope.
4. Determine the total amount of detention storage. Compute the detention storage between the water level at the point when the emergency starts to flow and the water level when the emergency is flowing at the design depth. If a principal spillway is used, add to the above amount the detention storage computed for the principal spillway. The total detention storage is V_s.
5. Compute V_s/V_r.
6. Using the value computed for V_s/V_r, determine the value of Q_{oc}/Q_i from Table 9.15.
7. Solve for Q_{oe}, the expected flow through the emergency spillway, in cfs.

Computation: For determining the size of the emergency spillway—

1. The total structure is designed for a twenty-five-year frequency storm.
2. The peak rate of runoff expected from the watershed for this twenty-five-year frequency storm is—

$$Q_i = Q_t \times L \times I \times T \times S \times V \times C \times P \times F$$
$$Q_i = 356 \times 1.04 \times 0.9 \times 0.96 \times 1.0 \times 0.88 \times 0.99 \times 0.98 \times 1.3$$
$$Q_i = 355 \text{ cfs}$$

3. The volume of runoff expected from the twenty-five-year frequency storm is obtained from Table 9.13. The product of $V \times I$ (0.88 × 0.9) = 0.79. The volume of runoff, acre-feet per acre = 0.16. Correcting for watershed location (location factor is 1.04) 0.16 × 1.04 = 0.166 acre-feet per acre. The total volume of runoff $V_r = 0.166 \times 160 = 26.6$ acre-feet.

TABLE 9.15. Relationship of Q_{oe}/Q_i to V_s/V_r for Emergency Spillway Design

V_s/V_r	Q_{oe}/Q_i									
	0.00	0.01	0.02	0.03	0.04	0.05	0.06	0.07	0.08	0.09
(a) All flow carried by emergency spillway										
0.2	0.91	0.90	0.89	0.87	0.86	0.85	0.83	0.82	0.80	0.79
0.3	0.77	0.76	0.75	0.73	0.72	0.71	0.69	0.68	0.66	0.65
0.4	0.64	0.62	0.61	0.59	0.58	0.57	0.55	0.54	0.53	0.52
0.5	0.51	0.49	0.48	0.47	0.46	0.44	0.43	0.41	0.40	0.39
0.6	0.39	0.37	0.36	0.35	0.34	0.32	0.31	0.29	0.28	0.27
0.7	0.28	0.26	0.25	0.24	0.23	0.21	0.20	0.18	0.17	0.16
0.8	0.17	0.16	0.15	0.14	0.13	0.11	0.10	0.09	0.08	0.07
0.9	0.07	0.06	0.05	0.04	0.03	0.02	0.01	0.01	0.00	0.00
(b) $Q_{op}/Q_i = 0.10$ (Capacity of principal equals 10% of peak rate of inflow)										
0.2	0.83	0.81	0.80	0.78	0.77	0.76	0.74	0.73	0.72	0.70
0.3	0.69	0.68	0.67	0.65	0.64	0.63	0.61	0.60	0.59	0.57
0.4	0.56	0.54	0.53	0.51	0.50	0.49	0.47	0.46	0.45	0.44
0.5	0.43	0.41	0.40	0.39	0.37	0.36	0.35	0.34	0.33	0.32
0.6	0.31	0.29	0.28	0.27	0.26	0.24	0.23	0.22	0.21	0.20
0.7	0.19	0.18	0.17	0.16	0.15	0.14	0.13	0.12	0.11	0.10
0.8	0.09	0.08	0.07	0.06	0.05	0.04	0.03	0.02	0.01	0.00
(c) $Q_{op}/Q_i = 0.20$ (Capacity of principal equals 20% of peak rate of inflow)										
0.2	0.75	0.74	0.72	0.71	0.69	0.68	0.66	0.65	0.63	0.62
0.3	0.61	0.60	0.58	0.57	0.55	0.54	0.52	0.51	0.50	0.49
0.4	0.48	0.47	0.45	0.44	0.42	0.41	0.39	0.38	0.37	0.36
0.5	0.35	0.34	0.32	0.31	0.30	0.29	0.27	0.26	0.25	0.24
0.6	0.23	0.22	0.20	0.19	0.18	0.17	0.15	0.14	0.13	0.12
0.7	0.11	0.10	0.08	0.07	0.06	0.05	0.04	0.03	0.02	0.01
(d) $Q_{op}/Q_i = 0.30$ (Capacity of principal equals 30% of peak rate of inflow)										
0.2	0.67	0.66	0.64	0.63	0.61	0.60	0.58	0.57	0.55	0.54
0.3	0.53	0.52	0.50	0.49	0.47	0.46	0.44	0.43	0.41	0.40
0.4	0.39	0.38	0.36	0.35	0.33	0.32	0.31	0.30	0.28	0.27
0.5	0.26	0.25	0.23	0.22	0.20	0.19	0.18	0.17	0.15	0.14
0.6	0.13	0.12	0.11	0.10	0.09	0.08	0.07	0.06	0.05	0.04
0.7	0.03	0.03	0.02	0.02	0.01	0.01	0.00	0.00	0.00	0.00

4. The amount of detention storage above the emergency spillway level is equal to the average pool area times the depth.

$$\frac{3.0 \text{ acres} + 5.6 \text{ acres}}{2} \times 1.8 = 7.7 \text{ acre-feet.}$$

Add the detention storage computed for the principal spillway, 5.0 acre-feet, to obtain the total detention storage V_s.

$$V_s = 7.7 + 5.0 = 12.7 \text{ acre-feet}$$

5. $V_s/V_r = 12.7/26.6 = 0.48$
6. The flow through the principal spillway, 52 cfs, is equal to 0.15 of the peak rate of runoff from the watershed, 355 cfs. $Q_{op}/Q_i = 52/355 = 0.15$. By interpolating between Tables 9.15(b) and 9.15(c) we find $Q_{eo}/Q_i = 0.41$.

7. $Q_{oe} = 0.41 \times 355 = 146$ cfs
8. The emergency spillway must have the capacity to carry 146 cfs at a depth of flow of 1.8 feet. By referring to Table 9.11, we find that the control section of the emergency spillway must be 23 feet wide.

Other depths of flow through the emergency spillway could be assumed with corresponding changes in the detention storage, and the above procedure repeated to determine the width of emergency spillway required. The combination that gives the most economical and most satisfactory structure is selected.

DETENTION STORAGE REQUIRED
FOR A SPECIFIC PRINCIPAL SPILLWAY DISCHARGE

In some cases a given size pipe may be available for the principal spillway, and you will want to determine the amount of detention storage required if this pipe is used. For example, assume that an 18-inch corrugated metal pipe is used as the principal spillway in the above structure. The capacity of this pipe is 19 cfs.

$$Q_{op}/Q_{ip} = 19/137 = 0.14$$

Refer to Table 9.14 and determine the value of V_{sp}/V_{rp} corresponding to a value of 0.14 for Q_{op}/Q_{ip}. A value of 0.75 is obtained. $V_{sp}/V_{rp} = 0.75$. The volume of runoff from the watershed, V_{rp}, was determined to be equal to 9.9 acre-feet. The detention storage required, $V_{sp} = 0.75 \times 9.9 = 7.4$ acre-feet. This additional detention storage can be obtained by increasing the difference in elevation between the principal and emergency spillways, which will require an additional height of embankment. The emergency spillway should be redesigned using this new value for detention storage.

TRICKLE TUBE

In some structures a principal spillway is not required to carry the peak rate of runoff. However, it may be necessary to lower the water level in the reservoir between runoff periods so that prolonged flow does not occur through the emergency spillway. A small trickle tube can be used for this purpose.

Example: The following example is used to illustrate when a trickle tube can be used and how to select the size. Assume the following watershed and structure characteristics:

1. Drainage area—80 acres (40 acres—land slope, 0.5 to 1.0 percent; 40 acres—land slope, 3 percent).
2. Location—central Missouri; $L = 1.0$.

3. Soil—Mexico silt loam; $I = 1.1$.
4. Vegetation—poor quality pasture; $V = 0.8$.
5. The emergency spillway conditions will allow it to be used once each half-year.
6. The reservoir pool will cover 3.5 acres when flow starts through the trickle tube.
7. The emergency spillway will be located 1 foot above the level of the tube.
8. The reservoir pool will cover 4.5 acres when flow starts through the emergency spillway.
9. Springs and seeps on the watershed contribute a flow of approximately 0.4 cfs during wet seasons.
10. The trickle tube will be 70 feet of smooth iron pipe. The total head causing flow would be 12 feet if the tube were flowing full.

Computation: To determine the size of trickle tube—

1. The volume of runoff expected from a half-year frequency storm, with a product of $V \times I$ $(0.8 \times 1.1) = 0.88$ and corrected for watershed slope, is computed as follows: 40 acres yielding (0.05–0.02) or 0.03 acre-feet per acre $= 1.2$ acre-feet; 40 acres yielding 0.05 acre-feet per acre $= 2.0$ acre-feet. The total runoff from the watershed is 3.2 acre-feet.
2. The volume of detention storage between the trickle tube and the emergency spillway will be $(3.5 + 4.5)/2 \times 1 = 4$ acre-feet.
3. In this case, the volume of detention storage is greater than the volume of runoff expected from a half-year storm. If the water level in the reservoir were lowered to the level of the trickle tube between runoff periods, flow could be expected through the emergency less frequently than once each half-year.
4. The trickle tube should not be less than 6 inches in diameter and should have the capacity to carry any prolonged flow from the drainage area and to lower the water to the level of the trickle tube in a reasonable length of time.

A convenient rule of thumb that can be used in selecting the size of the trickle tube is—*A tube capacity of 1 cfs will remove 1 inch of runoff from 1 acre (1 acre-inch) in one hour.*

Let us assume that an 8-inch smooth pipe with a capacity of 4.0 cfs (Table 9.5) when flowing full is to be considered for the trickle tube. Since the water level will not be high enough to cause full pipe flow throughout the drawdown period, a tube capacity approximately 40 percent of that given in Table 9.5 should be used. $0.40 \times 4.0 = 1.6$ cfs. A capacity of 0.4 cfs is required to carry the prolonged flow from the watershed. The remaining capacity, 1.2 cfs, will be available to lower the water level in the reservoir.

The volume of water stored between the trickle tube and the emergency spillway is 4 acre-feet or 48 acre-inches. Using the rule of thumb given above, we find that the 1.2 cfs will remove 1.2 acre-inches in one hour. So forty hours will be required to lower the water level in the reservoir to the level of the trickle tube.

PROBLEMS

The watershed shown in Figure P5.1 is located in the southeast corner of Kansas. The soil is a silt loam with a dense subsoil within 12 inches of the surface. Both surface and subsoil are of rather high plasticity.

9.1. An earth embankment structure is to be built at point A in Figure P5.1. The height of the embankment, measured from the lowest point along the center line to the top of the proposed fill, is 12 feet. The top of the proposed fill is to be 2.5 feet above the crest of the emergency spillway. The surface area of the reservoir at the emergency spillway level is 2.5 acres and the average depth is 7 feet. The following peak rates of runoff from the watershed can be expected: ten-year frequency storm—148 cfs, twenty-five-year frequency storm— 192 cfs, and fifty-year frequency storm—222 cfs.
   a. Give the frequency of storm for which the spillways should be designed.
   b. Assume that the principal spillway conduit is to be a corrugated metal pipe 60 feet long and is to carry 9 cfs. The total head causing flow is 11 feet. The difference in elevation between the upper and lower ends of the conduit is 10 feet. (1) Give the required diameter of conduit in inches. (2) Make a sketch showing all the dimensions needed to construct a drop inlet, a canopy inlet, and a hood inlet for this structure. Indicate the number, size, and location of antiseep collars.

9.2. A structure is to be built at point B in Figure P5.1 to lower the water from the waterway into the natural depression. The difference in elevation between the waterway and the bottom of the natural depression is 12 feet. The bank of the natural drainageway has a stable slope, has a good growth of fescue, buckbrush, and blackberry vines on the bank and extending back to the waterway, and the topography is such that the water tends to spread before flowing over the bank.
   a. What type of structure would you select if the natural drainageway is tending to cut deeper and a crossing is needed across the natural depression in this area?
   b. What type of structure would you select if a crossing is not needed?
   c. What additional types of structures may be considered if the natural drainageway is quite stable and only 6 feet deep?
   d. Design a structure for this location similar to the one shown in Figure 9.20. Assume that the pipe spillway is to be a smooth metal pipe 50 feet long, with a canopy inlet. Make a sketch showing all dimensions needed to construct the pipe spillway and the emergency spillway. Indicate the number, size, and location of anti-seep collars. The following peak rates of runoff can be expected from the 30-acre watershed: half-year frequency storm— 15 cfs, one-year frequency storm—23 cfs, ten-year frequency storm—76 cfs, and twenty-five-year frequency storm—99 cfs.
   e. Assume that the natural drainageway is quite stable and only 6 feet deep. What size straight drop spillway would you recommend? Would you use an emergency spillway to carry part of the runoff? If so, indicate the capacity, size, and location.

9.3. The principal spillway in the earth embankment structure at A (Fig. P5.1) is to be a corrugated metal pipe, 60 feet long. The total head causing flow is 11 feet. Assume that the reservoir has 2.2 acres of surface area at the principal spillway level and 2.5 acres of surface area at the emergency spillway level, 2 feet above the level of the principal spillway.

    a. Determine the size of pipe required to carry the expected flow from the reservoir resulting from a two-year frequency storm. Use $Q_{tp} = 74$ cfs.

    b. Determine the width of the control section required to carry the expected flow from the reservoir resulting from a twenty-five-year frequency storm. Use $Q_t = 192$ cfs. Assume a 1.3-foot depth of flow through the control section. The surface area of the reservoir at this level is 2.9 acres.

    c. The exit section is to be designed as a trapezoidal-shaped grassed waterway. The soil is less erodible and a good vegetative cover will be maintained. Determine the width and depth of grassed waterway required.

9.4. The principal spillway in the earth embankment structure at A (Fig. P5.1) is to be a smooth metal pipe 10 inches in diameter. The capacity of this pipe when flowing full is 7.0 cfs.

    a. How much detention storage below the emergency spillway level will be required if this pipe is to carry the two-year frequency runoff from the reservoir?

    b. What will be the width of the control section? Assume the same depth of flow and detention storage above the emergency spillway level as in Problem 9.3(b).

    c. What period of time will be required to lower the water level in the reservoir from the emergency spillway to the principal spillway level?

9.5. A principal spillway will not be used in the earth embankment structure at A (Fig. P5.1). All the runoff will be carried by the emergency spillway. Assume the same depth of flow through the control section and the same detention storage as in Problem 9.3(b).

    a. What will be the approximate width of the control section?

    b. What will be the approximate width of a grassed waterway exit section?

    c. Would you recommend this solution?

REFERENCES

1. Beasley, R. P., and Meyer, L. D. Hydraulic tests of erosion control structures—Sliced-inlet type entrance. Univ. Mo. Agr. Expt. Sta. Res. Bull. 599, January 1956.
2. Beasley, R. P., Meyer, L. D., and Smerdon, E. T. Canopy inlet for closed conduits. *Agr. Eng.*, vol. 41, no. 4, April 1960.
3. Culp, M. M. The effect of spillway storage on the design of upstream reservoirs. *Agr. Eng.*, vol. 29, no. 8, August 1948.
4. Schwab, G. O., Frevert, R. K., Edminster, T. W., and Barnes, K. K. *Soil and water conservation engineering.* New York: Wiley, 1966.
5. U.S. Soil Conservation Service, USDA. Engineering field manual for conservation practices, 1969.

10 ~ Farm Ponds

The primary function of a farm pond is to provide a water supply for livestock, wildlife, recreation, fire protection, and other uses. In building a good farm pond some factors must be considered that are not emphasized when building other types of soil and water conservation structures.

FARM PONDS fall into two main classifications, depending upon their construction characteristics: *embankment ponds* and *excavated ponds.*

Embankment, or impounding, ponds which impound a large portion of the water above the original ground surface are usually built in areas where the land slopes are gentle to moderately steep and a dam can be constructed across a depression. Surface water usually fills the pond.

Excavated, or dug-out, ponds have a large portion of the water stored below the original ground level. They are usually constructed where the land slopes are relatively flat and no site is available for an embankment pond. The fact that most of their capacity is secured by excavation limits their use to locations where a small supply of water is required. The pond may be filled by either surface runoff or groundwater seeping into the excavation.

In many ponds the capacity is obtained by both excavation and impoundment. For purposes of classification they are considered to be embankment ponds if the depth of the water stored above the original ground surface at the embankment exceeds 3 feet.

LOCATION

The selection of a suitable site for a pond is very important, and careful study of all possibilities should be made. Following are some of the factors to be considered in selecting the location:

233

Drainage area. The size of the drainage area is of primary importance. Where surface runoff is the main source of water, it is important that the contributing drainage area be large enough to maintain sufficient water in the pond. However, if the drainage area is too large, an excessive amount of water will flow into the pond and out through the spillway which may result in excessive siltation, and less desirable conditions for fish production, and may necessitate a larger and more expensive spillway. The size of the drainage area may be changed by diverting the desired amount of water into or out of the pond drainage area by the use of terraces or diversions. It may be increased by locating the pond beside a waterway at such a level that when the pond is not full, water from the waterway will flow into the pond, and when the pond is full, the majority of the water will not enter the pond but will flow on down the waterway.

Erosion of the drainage area contributes sediment to the pond, thus reducing its capacity and desirability. It is advantageous to have the entire drainage area in permanent grass or timber. If sloping land is to be cultivated, practices which will reduce erosion to a minimum are employed. If runoff from eroding land belonging to another owner enters the pond, a debris basin is constructed above the pond to collect the sediment.

Pollution of the pond water should be avoided by selecting a site where drainage from feedlots, sewage lines, mine dumps, and similar areas will not reach it. Diversions can sometimes be used to keep undesirable drainage from entering the pond. It is desirable to locate the pond so that all drainage into it comes from land belonging to the owner. He will thus have control of the drainage area.

Capacity and economics of storage. The topography of the site should be such that a pond of the desired capacity can be obtained at a reasonable cost. From an economic standpoint, the pond should be located where the largest storage volume can be obtained with the least amount of earthmoving. This condition will generally occur where a dam constructed across a narrow drainageway will back water over a large area.

Convenience. Livestock water ponds are located to serve a large grazing area with a minimum of travel by the animals. If the pond is to be used as a source of water for the farmstead, it should be located nearby, preferably at a higher elevation. For fire protection purposes the pond should be located within 500 feet of the farm buildings and near an all-weather road.

An attempt should be made to locate the pond so it can serve more than one use. Most ponds can be used for fishing and recreation and as wildlife refuges, in addition to other uses.

Soils. The success of a pond is dependent on the ability of the soils in the reservoir area to hold water and to provide a stable embankment. In areas where the nature of the soil is such that permeable or unstable materials may be present, soil borings are taken prior to construction to determine if a satisfactory structure can be built at that site.

Shallow areas. A site should be selected that will not have large shallow areas, since most weeds which result in undesirable conditions start in shallow water. Elimination of shallow water along the edges of the pond during construction is accomplished by filling low areas with earth so that water will not be impounded on them, and by grading down from the waterline on a three-to-one slope to a depth of at least 3 feet.

Obstructions. Buried pipelines or cables may be damaged during construction, and overhead power lines present a hazard to pond fishermen.

DESIGN

STORAGE CAPACITY

The farm pond must have adequate capacity to provide water which will be (1) used from the pond, (2) lost by evaporation and seepage, and (3) retained in a reserve pool. Most areas are subject to drought periods when no surface runoff occurs to replenish the supply. It is important that the capacity be adequate to supply the water that will be used or lost from the pond during the longest period with inadequate runoff to be expected in the area.

Use. Household use requires 50 gallons per day per person; beef cattle and horses use 15 gallons per animal per day; dairy cattle require 20 gallons per animal per day for drinking only, and another 30 gallons per animal per day are required for barn and utensil cleanup; hogs use 5 gallons per animal per day; and 100 chickens require 10 gallons per day.

Evaporation. In a dry year the amount of water evaporating from the surface of a pond may exceed the amount replaced by rainfall on the pond. The excess of evaporation over rainfall can be obtained locally from the National Weather Service. The evaporation records obtained from them may be the evaporation from Class A pans. If so, these values should be multiplied by 0.7 to obtain the evaporation from a pond. The percent reduction in capacity by evaporation from typical farm ponds can be determined from Table 10.1.

Seepage. Losses vary widely depending upon the soil and the practices followed during construction. If practices suggested in Chapter 9 are followed, seepage losses should be negligible for ponds constructed in less permeable soils and should not exceed 5 percent of the total volume per year for ponds located on more permeable soils.

Reserve pool. Retaining a certain amount of water in a reserve pool will prevent the pond dam from drying out and cracking and will provide water for wildlife and recreation and an area for sediment ac-

TABLE 10.1. Percent Reduction in the Capacity of Typical Farm Ponds for Each 1.0 Foot Net Loss by Evaporation

Extent to Which Water is Depleted during Period	Maximum Depth of Water in Pond (ft)		
	8	10	12
Water maintained at near full reservoir during the period	22	18	15
Entire supply of water depleted during the period	13	10	8
One-half the maximum depth used during the period	18	14	12
All but 4 feet of water used during the period	18	13	11

cumulation. The depth of water in the reserve pool will vary depending on the desired use and the expected sediment accumulation. The percentage of the total volume of the pond in a reserve pool can be determined from Table 10.2.

Example: A pond, 10 feet deep, is to be constructed in an area where local National Weather Service records indicate that evaporation will exceed precipitation by 1.2 feet in a dry year. The pond is to provide drinking water and water for barn and utensil cleanup for fifty dairy cows in production, drinking water only for twenty-five beef cattle, household water for a family of five people, and a reserve pool 5 feet deep.

Computation: Water requirements for intended uses—

50 producing cows @ 50 gallons	=	2,500 gallons per day
25 beef cattle @ 15 gallons	=	375 gallons per day
5 people @ 50 gallons	=	250 gallons per day
Total		3,125 gallons per day

TABLE 10.2. Percentage of the Total Volume of Typical Ponds with Varying Depths of Water in the Reserve Pool

Depth of Water in Reserve Pool (ft)	Total Depth of Pond (ft)		
	8	10	12
12	. . .	. . .	100
11	. . .	. . .	85
10	. . .	100	71
9	. . .	82	58
8	100	66	47
7	78	51	37
6	58	39	28
5	42	29	21
4	28	20	15
3	18	13	11
2	10	8	7
1	5	4	3

The pond is to be built large enough to provide sufficient water to supply the above uses for a one-year period. The yearly use will be $365 \times 3,125 = 1,140,625$ gallons.

An acre-foot, which contains 326,000 gallons, is a more convenient unit of volume to use. In the above example, the yearly use will be $1,140,625 \div 326,000 = 3.5$ acre-feet.

Evaporation Losses. Evaporation from the pond surface in a dry year will exceed precipitation by 1.2 feet. The evaporation losses will equal 1.2 times 14, or 17 percent of the total volume if the water is not withdrawn below the 5-foot level (Table 10.1).

Seepage. In this example assume the seepage losses to be 2 percent.

Reserve Pool. The 5 feet of water retained in the reserve pool will amount to 29 percent of the total volume (Table 10.2).

Total losses.

$$
\begin{array}{ll}
\text{Evaporation} & = 17 \text{ percent of the total volume} \\
\text{Seepage} & = 2 \text{ percent of the total volume} \\
\text{Reserve pool} & = \underline{29} \text{ percent of the total volume} \\
\text{Total} & 48 \text{ percent of the total volume}
\end{array}
$$

Total Storage Capacity. Since the losses amount to 48 percent of the total volume, the amount of water needed for the intended uses (3.50 acre-feet) equals 52 percent of the total capacity.

$$
\begin{array}{l}
52 \text{ percent} = 3.5 \text{ acre-feet} \\
100 \text{ percent} = 6.73 \text{ acre-feet} = \text{total capacity of the pond}
\end{array}
$$

SURFACE AREA AND DEPTH

During long dry seasons the depth of water evaporated from the pond surface is appreciable. The pond should be constructed to such depth that, after evaporation losses occur, sufficient water will remain to supply the demand. The minimum depth should be 6 feet in humid areas, 8 feet in subhumid areas, and 12 feet in semiarid areas. The minimum depth of water should extend over an area equal to at least one-sixteenth and, if possible, up to one-fourth of the surface area.

In laying out a farm pond, it is necessary to select the water surface area and the depth that will give the capacity needed in the pond. The following approximate relationships between surface area, depth, and capacity for typical ponds may be helpful in making this determination:

1. For small ponds where the excavation is made over the entire area covered by the water

$$
\text{Surface area (acres)} = \frac{\text{storage (acre-feet)}}{0.5 \text{ times the maximum depth (feet)}}
$$

2. For large ponds where the excavation is not made over the entire area covered by the water

$$\text{Surface area (acres)} = \frac{\text{storage (acre-feet)}}{0.4 \text{ times the maximum depth (feet)}}$$

If a more accurate determination of the capacity of a pond and its relation to the surface area is needed prior to construction, the following procedure is suggested:

1. Prepare an accurate contour map of the area on which the pond is to be built. Select the contour which represents the water level in the pond.
2. Draw the proposed fill on this map, indicating where the fill intercepts the contour lines.
3. The volume of water stored above the original ground surface can be computed by determining the average area of adjacent contours, multiplying each average area by the vertical distance between contours, and finding the sum of these products.
4. Determine the volume of fill material which will be excavated from below the water level contour and add to the volume of water stored above the original ground surface to obtain the total capacity of the pond.

SIZE OF DRAINAGE AREA

In estimating the size of drainage area required to maintain a reliable supply of water in the pond, the amount of runoff expected for the period must be determined. The most critical period will occur when rainfall is limited, and usually this will be a period of high evaporation losses.

The amount of runoff from a watershed is dependent on many variables. The physical characteristics of the watershed that effect runoff are land slope, soil infiltration, vegetative cover, and surface storage. Rainfall characteristics such as amount, intensity, and duration vary widely throughout the nation and from season to season. Considering all the variables, it is obvious that any set rule given for the determination of watershed size may be subject to appreciable error for a given pond during a certain period of time.

The values on the map in Figure 10.1 are given to serve as a general guide for estimating the size of drainage area required per acre-foot of water used or lost by evaporation and seepage from the pond in a dry year. These values assume that the pond is large enough to provide sufficient water for a one-year period. If the pond is constructed large enough to provide water for a two-year period, the size of the drainage area re-

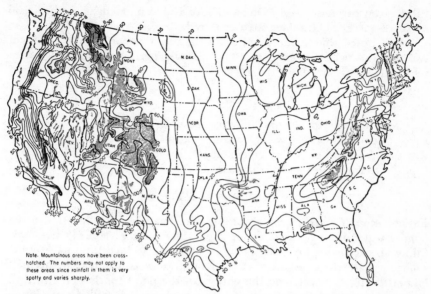

Figure 10.1. Guide for estimating approximate number of acres of drainage area required per acre-foot of water used or lost by evaporation and seepage from the pond in a dry year. Reproduced from U.S. Soil Conservation Service, USDA, Engineering field manual for conservation practices, 1969.

quired per acre-foot of water used or lost by evaporation and seepage will be one-fourth the size given in Figure 10.1. The drainage area can be reduced since the runoff to be expected in two consecutive years is approximately four times the amount expected in a dry year.

Example: In the previous example, a storage capacity of 6.73 acre-feet was required. Of this total volume, 52 percent was for the intended use, 17 percent was for evaporation losses, 2 percent was for seepage losses, and 29 percent was in the reserve pool.

Since water in the reserve pool is not used, it is only necessary to replace the water which is used and that which is lost by evaporation and seepage, which amounts to 71 percent of the total volume. Seventy-one percent of 6.73 equals 4.78 acre-feet.

This pond is located in western Missouri. From Figure 10.1, it is determined that 10 acres of drainage area will be required per acre-foot of water to be replaced during the one-year period of intended use. Thus 10 times 4.78 equals 48 acres, the size of drainage area required. If the pond were built large enough to provide a reliable supply of water for a two-year period, the size of the drainage area could be reduced.

In the above example, the water needed for the intended use plus evaporation and seepage losses for a two-year period would be 2 times 4.78, or 9.56 acre-feet.

The drainage area per acre-foot of storage will be one-fourth the value given in Figure 10.1. One-fourth of 10 is 2.5. Hence 2.5 times 9.56 equals 24 acres of drainage area required.

It is noted that if the pond size is increased to provide water for a two-year use period instead of a one-year use period, the drainage area required is only one-half as large.

SPILLWAY

The spillway for a pond may consist of principal spillway, a trickle tube, an emergency spillway, or a combination of either a principal spillway or trickle tube and an emergency spillway. These should have the capacity to carry the design storm (see Ch. 9).

WATER SUPPLY SYSTEMS

Embankment Ponds. A water supply pipe may be installed under the dam if water is to be used at some point below for watering livestock, filling an orchard spray tank, or other uses. A 1¼-inch pipe will supply sufficient water for most uses. A filter or strainer is installed at the inlet end of the pipe. A filter on the bottom of the pond consists of a vertical section of pipe with four rows of ¼-inch holes drilled on 3-inch centers surrounded by a perforated container filled with crushed stone or pea gravel. For best quality water, a floating or suspended strainer or filter is used which takes water from near the surface of the pond. A cutoff valve is installed in the pipe near the back toe of the dam. The valve has a vertical enclosure to provide accessibility and to prevent freezing.

Excavated Ponds. Water cannot be withdrawn from lower levels of an excavated pond through an underground pipe by gravity. A ramp with a slope of four to one or flatter is constructed for livestock access.

CONSTRUCTION

Embankment ponds are constructed by the same methods utilized for the construction of earth embankments (Ch. 9).

Although excavated ponds may be constructed in almost any shape, a rectangular-shaped pond is usually the most convenient for excavating equipment. Side slopes of two to one are commonly used, except for the access ramp which has a slope of four to one or flatter. The excavated material may be used to build ridges to divert additional runoff into the pond. If the pond is located on sloping land, excavated material may be used in an embankment around the lower portion of the pond to increase its capacity.

SEALING PONDS AND RESERVOIRS

Much of this information on sealing ponds and reservoirs was taken from U.S. Soil Conservation Service (1).

Excessive seepage losses in farm ponds are usually due to the selection of a site where the soils are too permeable to hold water. This may be the result of inadequate site investigations in the planning stage. However, the need for water may be so important as to justify the selection of a permeable site. In such cases, plans for reducing seepage losses by sealing should be part of the design.

The problem of reducing seepage is one of reducing the permeability of the soils to a point where the losses become tolerable. Losses may be reduced by the methods discussed below, the choice of which will depend largely on the proportions of coarse-grained sand and gravel and fine-grained silt and clay in the soil. A thorough investigation of the materials to be sealed should be made by a soils scientist before the method of sealing is selected. In some cases it may be necessary to have a laboratory analysis of the materials.

In all methods of treatment, the pond area should be cleared of trees, brush, stumps, roots, and other debris, and holes and crevices filled with compacted material prior to the treatment.

Compaction. Pond areas containing a high percentage of coarse-grained material can be made relatively impervious by compaction alone if the material is well graded from small gravel or coarse sand to fine sand, silt, and clay. This method of sealing is the least expensive of those presented, but its use is limited to the soil conditions described.

The soil should be scarified to a depth of 8 to 10 inches with a disk, roto-tiller, pulverizer, or similar equipment and all rocks and tree roots should be removed. The loosened soil should be rolled, under optimum moisture conditions, to a dense, tight layer with four to six passes of a sheepsfoot roller.

The thickness of the compacted seal should not be less than 8 inches for impoundments up to 10 feet in depth. Since seepage losses vary directly with the depth of water impounded, the thickness of the compacted seal should be increased by 1 inch for each foot of water over 10 feet. This will require compacting the soil in two or more layers, not exceeding 8 inches in thickness, over that portion of the pond where the water depth exceeds 10 feet. In these cases, it will be necessary to remove the top layer or layers of soil and stockpile it while the bottom layer is being compacted.

Clay Blankets. Pond areas containing high percentages of coarse-grained soils, but lacking sufficient amounts of clay to prevent high seepage losses by compaction alone, can be sealed by blanketing. The blanket should cover the entire area over which water is to be impounded. It should consist of material containing a wide range of particle sizes varying from small gravel or coarse sand to fine sand and clay in the desired proportions. Such material should contain approximately 20 percent by weight of clay particles.

The thickness of the blanket will depend on the depth of water to be impounded. The minimum thickness should be 12 inches for all depths of water up to 10 feet. The minimum should be increased by 2 inches for each foot of water over 10 feet.

Bentonite. Seepage losses in well-graded coarse-grained soils may be reduced by the addition of bentonite. Bentonite is a fine-textured colloidal clay that will absorb several times its own weight of water. At complete saturation, it will swell from eight to fifteen times its original volume. When bentonite is mixed in the correct proportions with the coarse-grained material and the mixture is thoroughly compacted and saturated, the particles of bentonite will fill the pores in the material and make it nearly impervious. A laboratory analysis of the material is essential to determine the amount of bentonite that should be applied per unit of area. Rates of application range from 1 to 3 pounds per square foot, depending on site conditions. Upon drying, bentonite will return to its original volume and leave cracks in the pond area. For this reason bentonite is not recommended for ponds where a wide fluctuation in the water level is expected.

For good compaction, the soil should be moist. If the material is too dry, water should be added by sprinkling. However, if the area is too wet, sealing operations should be postponed until moisture conditions are satisfactory. The bentonite should be spread uniformly over the area to be treated at the rate determined by the laboratory analysis. The bentonite is thoroughly mixed with the soil to a depth of at least 6 inches with a roto-tiller, disk, or similar equipment. The area should then be compacted with four to six passes of a sheepsfoot roller.

Chemical Additives. Excessive seepage losses often occur in fine-grained clay soils because of the arrangement of the clay particles which form a honeycomb structure. Such soils are said to be aggregated and have a relatively high permeability rate. The application of small amounts of certain chemicals to these aggregates may result in collapse of the open structure and rearrangement of the clay particles. The chemicals used are called dispersing agents.

For chemical treatment to be effective, the soils in the pond area should contain more than 50 percent of fine-grained material (silt and clay finer than .074 mm diameter) and at least 15 percent of clay finer than .002 mm diameter. The soils should contain less than 0.5 percent soluble salts (based on dry soil weight). Chemical treatment is not effective in coarse-grained soils.

While there are many soluble salts that meet the requirement of a dispersing agent, sodium polyphosphates are most commonly used. Tetrasodium pyrophosphate (TSPP) and sodium tripolyphosphate (STPP) are most effective. These dispersants should be quite fine with 95 percent passing a no. 30 sieve and less than 5 percent passing a no. 100 sieve. They are usually applied at a rate of from 0.05 to 0.10 pounds per square foot. Sodium chloride, which is less effective, is applied at a rate of from 0.20 to 0.33 pounds per square foot. A laboratory analysis of the soils in the pond area is essential to determine which of these dispersing agents will be most effective and the rate of application.

The dispersing agent is mixed with the surface soil and compacted to form a blanket. For depths of water up to 8 feet the banket thickness should be at least 6 inches. For depths of water greater than 8 feet the blanket should be 12 inches thick and treated in two 6-inch lifts.

The soil moisture level in the area to be treated should be near optimum for compaction to a depth of 12 inches. If the soil is too wet, treatment should be postponed. Polyphosphates release water from the soil, and the job could easily become too wet to handle. If the soil is too dry, water should be added by sprinkling.

The dispersing agent should be applied uniformly over the pond area at a rate determined by the laboratory analysis. It may be applied with a seeder, drill, fertilizer spreader, or by hand broadcasting, and should be thoroughly mixed into each 6-inch layer with a disk, rototiller, pulverizer, or similar equipment. Operating the mixing equipment in two directions will produce best results. Each chemically treated layer should be thoroughly compacted with four to six passes of a sheepsfoot roller. The treated blanket should be protected from puncture by livestock trampling.

Flexible Membranes. Another method of reducing excessive seepage losses is the use of flexible membranes of polyethylene, vinyl, or butyl rubber. Thin films of these materials are structurally weak but, if kept intact, they are almost completely watertight. Polyethylene films are less expensive and have better aging properties than vinyl. Vinyl is more resistant to impact damage and is readily seamed and patched with a solvent cement. Polyethylene can be joined or patched only by heat sealing. Butyl rubber can be joined or patched with a special cement.

The area to be lined should be drained and allowed to dry until the surface is firm and will support the men and equipment that must travel over it during installation of the lining.

Certain plants penetrate vinyl and polyethylene film. For this reason, it is desirable to sterilize the subgrade with chemicals, particularly the side slopes where nut grass, Johnson grass, quack grass, and other plants having a high penetrating power are present. Sterilization is not required where butyl rubber membranes are used.

The top edges of the lining should be anchored in a trench excavated completely around the area to be lined. The trench should be 8 to 10 inches deep and about 12 inches wide. The lining should then be anchored by burying 8 to 12 inches of the lining in the anchor trench and securing it with compacted backfill.

The linings are usually laid in sections or strips with a 6-inch overlap for seaming. Vinyl and butyl rubber linings should be laid smooth but in a loose state. Polyethylene should have up to 10 percent slack.

These thin films must be protected from mechanical damage if they are to be serviceable. All polyethylene and vinyl rubber membranes should have a cover of earth or earth and gravel not less than 6 inches thick. Butyl rubber membranes need to be covered only in areas subject to travel by livestock. In these areas a minimum cover of 9 inches should be used over all types of membranes. The materials used to cover the membrane should be free of large clods, sharp rocks, sticks, and other objects that would puncture the lining. The bottom 3 inches of cover should not be coarser than silty sand.

FENCING

A permanent fence around the pond is desirable to prevent livestock from damaging the embankment and spillway and muddying the water. The fence should be located at least 10 feet from the downstream toe of the dam and at least 40 feet from the water's edge. A temporary fence can be used across the ramp of excavated ponds to keep livestock out of the main part of the pond. However if livestock do not have access to the pond, a fence may not be required.

BUFFER STRIP

A strip of permanent vegetation at least 100 feet wide extending out from the waterline is desirable. This will improve the appearance of the pond and reduce siltation.

PROBLEMS

10.1. A pond 12 feet deep is to be constructed in the northeast corner of Kansas. Local National Weather Service records indicate that rainfall expected in a dry year is 29 inches and the evaporation from a class A pan in a dry year is 67 inches.

a. Give in inches the excess of evaporation over rainfall to be expected in a dry year.

b. The pond is to be built large enough to provide water for two hundred head of beef cattle with a reserve pool 4 feet deep. The seepage losses to be expected will be about 3 percent. If the pond were constructed to supply water for a one-year period, give the capacity in acre-feet.

c. In constructing the pond, excavation will be made throughout the area covered by the water. Give the approximate surface area of the pond in acres.

d. Give in acres the size of drainage area required to provide a reliable supply of water for the one-year use period.

e. If the size of the pond were increased to provide a reliable supply of water for a two-year use period, give the size of the drainage area in acres.

REFERENCE

1. U.S. Soil Conservation Service, USDA. Engineering field manual for conservation practices, 1969.

11 〜 Planning Agricultural Systems

It is important that production practices
be incorporated into a system that
will provide an acceptable income and
will reduce erosion to an acceptable level.

SOIL AND CROP management practices that are adopted should improve the infiltration and water-holding capacity and increase resistance to dispersion and movement by raindrop splash and runoff water. Soil management and fertility practices that increase production not only increase the income potential of the land, but increased vegetative growth results in decreased erosion.

An excellent growth of vegetation on the land is effective in reducing runoff and erosion in the following ways:

1. It intercepts a part or all of the precipitation. The vegetative canopy above the surface and residue on the surface intercept raindrops and absorb their impact.
2. Vegetation and vegetative residue on the surface reduce the velocity of the runoff, thus tending to prevent concentration of runoff water and increasing infiltration time.
3. Roots bind the soil particles together and when decayed add humus and provide openings for water to enter the soil.
4. Vegetation and its residue provide a suitable habitat in which small animals, earthworms, and other biotic life may function in the process of soil building. Their activity in the soil improves its physical condition and their channels and passageways make the soil more permeable.
5. Living plants take water from the soil in the process of growth, providing capacity for future rains.

The effectiveness of different types and amounts of vegetative cover is indicated by the values of the crop and management factor C in the soil-loss equation in Chapter 4. The factor C is the ratio of soil loss from

TABLE 11.1. Effect of Type and Amount of Vegetative
             Cover on Erosion

Type of Cover	Soil-Loss Ratio (%)
Fallow (bare)	100.0
Row crop	40.0
Small grain	10.0
Meadow	0.6

land under specified conditions to the corresponding loss from continuously fallow and tilled land. In this chaper, C is expressed as a
percentage of the expected loss from the soil in a fallow tilled condition.
This is designated the *soil-loss ratio*.

Type of cover. The type and amount of cover has a significant effect
on the expected soil loss (Table 11.1).

Stage of growth and amount of cover. Crops offer varying degrees
of protection against erosion depending upon their stage of growth.
Also, higher-producing crops resulting from improved soil management
practices provide a better crop canopy and return more residue, which
increases their effectiveness in reducing erosion. The soil-loss ratios for
corn at various levels of production and for the various crop stages defined in Chapter 4 are given in Table 11.2.

Previous year's crop. A good vegetative cover not only is effective in
controlling runoff and erosion while the crop is on the land but also
affects erosion during subsequent years. The soil-loss ratios for a corn
crop following a variety of crops are given in Table 11.3.

Management of crop residues. If the residue from a corn crop is
left in the field, the soil-loss ratio for stage 4 between harvest and plowing
will be 35. If the entire crop is removed for silage, the soil-loss ratio
for stage 4 will be 60. Residues are most effective when left on the surface; however, they do provide some control if they are worked into the
soil. Methods of seedbed preparation for row crops that leave a part or
all of the residue of the previous crop on the ground surface are effective
in reducing soil erosion. The runoff and soil loss resulting from a series
of high-intensity simulated storms on row crops with residue applied at
different rates as a mulch to the soil surface are given in Table 11.4 (8).

Although crop residues on the surface effectively reduce soil erosion,
there are disadvantages associated with their use. The residue on the

TABLE 11.2. Effect of Corn Yield on Erosion during Various Crop Stages

Yield (bu per acre)	Soil-Loss Ratios				
	Crop stage				
	F	1	2	3	4
Over 75	36	63	50	26	30
60–74	45	66	54	29	40
40–59	55	70	58	32	50
20–35	70	76	64	38	65

TABLE 11.3. Effect of Previous Crops on Erosion from Land Planted in Corn during Various Crop Stages

Previous Crop	Soil-Loss Ratios				
	Crop stage				
	F	1	2	3	4
Clover hay	21	35	32	25	35
Sweet clover	23	45	38	28	35
Lespedeza hay	55	70	60	32	50
Grass-legume meadow	15	32	30	19	30
Second year of corn following grass-legume meadow	42	57	49	28	42
Third year of corn following grass-legume meadow	55	70	58	32	50

surface will reduce the soil temperature. Research indicates that this reduction may be approximately 2° F during the early part of the growing season (12). This lower temperature delays germination, emergence, and early growth which may be a serious disadvantage in areas with a short growing season. Problems in placement of fertilizer for maximum nutrient efficiency and in weed, insect, and disease control will be different than those associated with conventional tillage methods.

The amount of residue produced varies widely depending upon the crop. Soybeans, for example, produce a limited amount of residue which decays quickly. Consequently, the residue from the soybean crop, even if left on the surface, will provide little protection from erosion.

TILLAGE

The ideal tillage method must provide for—

Seed germination. A seedbed that will provide the desired temperature, moisture, and aeration for the particular seed should be prepared (9). The seedbed should be firm, but not subject to crusting of the surface by raindrop splash or other sources of mechanical impedance to emergence.

Root growth and plant development. The physical characteristics of some soils provide ideal conditions for root development and plant growth at different stages of development. Other soils may have

TABLE 11.4. Effect of Different Amounts of Residue on Runoff and Erosion

Mulch Rate (tons per acre)	Runoff (in)	Soil Loss (tons per acre)
None	3.3	14.5
¼	2.8	5.8
½	2.4	3.7
1	2.0	1.7

Source: L. D. Meyer and J. V. Mannering. Tillage and land modifications for water erosion control. Am. Soc. Agr. Eng. Publ. Proc-168, 1967.

TABLE 11.5. Effect of Tillage Method on Cumulative Infiltration

Tillage Treatment	Cumulative Infiltration (in)	
	To initial runoff	To 2-inch runoff
Untilled	0.35	0.96
Plow	6.74	9.07
Plow-disk-harrow	2.10	3.32
Cultivated	2.23	3.59
Rotavated	0.94	1.63

Source: R. E. Burwell, R. R. Allmaras, and L. L. Sloneker. Structural alteration of soil surfaces by tillage and rainfall. *J. Soil Water Conserv.*, vol. 21, no. 2, January–February 1966.

layers in the soil profile that restrict root development and water movement. The water intake, water-holding capacity, soil aeration, and root development in these soils may be increased by deep tillage (3).

Weed control. Weeds must be destroyed by tillage, or the methods must be such that weed control can be accomplished by application of herbicides.

Increased infiltration and moisture-holding capacity of soil and reduction in evaporation losses. The tillage method can have a significant effect on the infiltration of water into a soil. Simulated rainfall was applied at a rate of 5 inches per hour to a Barnes loam soil on a 4 percent slope (10). The effect of the tillage treatments on cumulative infiltration is shown in Table 11.5 The effect of tillage practices on the water storage capacity of a soil is not as great as on the infiltration rate. Probably the greatest effect would be on varying the deph of the water storage zone (1). Tillage operations which leave the crop residue on the surface will reduce evaporation losses but will delay drying and reduce temperatures.

Control of undesirable insects or disease. Tillage practices that provide an environment favorable to destructive insects or disease should be avoided, or effective control methods should be developed.

Growth and activity of soil microorganisms. Tillage practices that mix plant residues with the soil may speed up the activity of microorganisms in decomposing crop residues and soil organic matter, with consequent increased availability of mineral nutrients, but with accompanying losses of nitrogen and organic matter. Minimum tillage practices will probably result in less microbial activity and less loss of nitrogen and organic matter and place more dependence on fertilization to supply the mineral nutrients. The effect on a specific microbial function such as nitrification, denitrification, or disease incidence may be of prime importance in evaluating a tillage program (7).

Proper placement of fertilizer, insecticides, and herbicides. The ideal fertilizer placement is influenced by the type of tillage and the weather conditions (4).

TABLE 11.6. Effect of Tillage Method on Runoff and Erosion

Tillage Method	Runoff (in)	Soil Loss (tons per acre)
Conventional	3.1	22.3
Wheel-track	2.3	17.0
Mulch tillage	2.5	6.7

Source: L. D. Meyer and J. V. Mannering. Tillage and land modifications for water erosion control. Am. Soc. Agr. Eng. Publ. Proc-168, 1967.

Timeliness and efficiency in planting and harvesting. With large acreages planted by single operators, it is important that the tillage method be adapted to use with large machinery so that even the large acreage can be planted during the most desirable planting period. The power required should be low, and a seedbed should be prepared with a minimum number of trips over the field. The method must also leave the ground in a condition that will not interfere with harvesting operations.

Protection from erosion. In order to provide maximum protection from erosion, a tillage method should minimize the amount of surface tillage and traffic, leave the surface rough and cloddy, make use of crops or crop residues for protecting the soil surface, and maximize the permeability of the subsurface profile.

Table 11.6 gives the runoff and soil loss resulting from a high-intensity simulated storm on corn with rows on a 6 percent slope under different tillage methods.

Methods. In an attempt to meet the above requirements, a number of different types of tillage have been used.

Conventional tillage. The land is plowed with a moldboard or disk plow and is given one or more disk harrowings followed by a spring or spike-tooth harrowing.

Plow-plant or wheel-track plant. Plowing with a moldboard or disk plow is the only operation prior to planting. In the plow-plant method, plowing and planting are done in a single operation. In the wheel-track method, planting is done in a second operation by planting in the wheel track left by the tractor.

Mulch tillage. The primary tillage is done with a chisel plow or sweeps, tending to leave the residue of the previous crop on the surface or to mix it with the surface layer of soil.

Rotary tillage. The ground is loosened and mixed by power-driven rotary blades. This is sometimes called *rotavating.*

Listing. Two furrows are turned laterally in opposite directions, thereby giving a ridge-and-furrow configuration. The seed is usually planted in the bottom of the furrows as the land is listed. When used on unplowed ground, it is called hard ground listing; on plowed ground it is called loose ground listing.

Deep tillage (subsoiling). A chisel or sweep is pulled through the soil at greater than normal plowing depth. This operation is usually performed in addition to the normal seedbed preparation.

Vertical mulching. A vertical band of mulching material is injected into the slit formed by deep tillage operation.

Strip tillage. Because of the difficulty of meeting all of the requirements of good tillage by one tillage method, the two-zone concept of tillage for row crops was conceived. In this method a narrow strip of soil is specifically prepared to meet the needs of the germinating seedling and the young plant, and the inter-row area is left in a condition which will be conducive to infiltration of rainfall and be resistant to erosion. The width of the prepared strip varies from a few inches to 1 foot or more, depending upon the equipment used. The strips may be prepared in the residue resulting from the harvest of the previous crop. A growing crop such as grasses or small grain may be killed by herbicides or excessive application of anhydrous ammonia on the strip prepared in the residue. Researchers are hopeful of developing a method that will stunt the growth of grass so it will not compete with the row crop for moisture and nutrients during the growing season, but will recover in the fall to provide winter cover as well as a cover in which the crop can be planted the following year.

The planting of row crops in sod by strip tillage is very effective in controlling erosion; however this is a relatively new practice and the management practices required in order for it to be widely adopted have not been developed.

Some scientists are exploring the possibility of using early maturing winter annuals to give protection during periods when row crops are susceptible to erosion. These winter annuals germinate in early fall, continue growth during the winter and spring, and mature in late spring or early summer. The row crop is planted in narrow strips between rows of the winter annual which matures and produces seed before offering competition to growth. The seed germinates in the fall to give the desired protection after the row crop is harvested.

Ridge planting. Row crops are planted on ridges and the residue of the previous crop accumulates in the furrow between ridges. Since the ridge is well drained and relatively free from residue, the soil warms up more quickly and the disadvantage of low soil temperature associated with residues is reduced. The residue that is intermixed with the soil in the furrow reduces erosion.

One system of ridge planting is to plant on the ridge left from the previous crop. The residue of the previous crop is windrowed into the furrows. This leaves a relatively trash-free surface at the center of the old ridge for planting. The ridge is reconstructed by disk hillers during later cultivations. In this system a ridge is provided for each crop row (2).

Another system uses larger, more or less permanent ridges constructed to a size and grade that will carry the runoff to a controlled outlet. They may be wide enough for a single crop row or for two or more (2, 5). The tractor wheels are operated on the ridges. Because of the accumulation of residue, the lack of traffic in the furrows, and the opportunity for unimpeded exhausting of displaced air through the ridges, infiltration is improved and erosion reduced. The use of ridges permits the planting of crop rows on a steeper grade than is possible with conventional tillage methods.

Obviously, no single tillage practice will meet the requirements for all crops on all soils in all climates. The combination of tillage and water management practice should be selected that best meets the requirement for a specific crop on a specific soil under the existing climatic conditions.

CONTOURING

The practice of performing tillage operations and planting crop rows across the slope is commonly called contouring or contour farming. Furrows, crop rows, and wheel tracks across the slope act as miniature terraces which detain the water, resulting in increased infiltration. The water that does not infiltrate into the soil is diverted to grassed waterways for disposal from the field. Since the capacity of these miniature terraces is limited, this practice is most successful on permeable soils and in areas of low-intensity rainfall where the surface runoff is limited. Also, the slopes must be quite uniform so that a grade can be maintained in the crop rows without the rows being too crooked, and so that an excessive number of correction areas is not required to keep the grade within design limits.

Contour farming conserves more moisture for crop growth, reduces the loss of stand due to the high velocity of runoff down crop rows planted up and down the slope, requires less power since farming operations will be nearly on the contour, and results in less soil erosion. Its effect on soil erosion is quite variable depending upon the soil, climate, and topography. The values given in Table 4.10 can be used to estimate the effectiveness of contour farming on soil erosion if information is not available for specific conditions in the area being considered.

Locating Guidelines. Guidelines should be located so that the grade in the crop rows will be within certain limits. The desired grade will vary depending on the expected amount of runoff and the erosiveness of the soil. On permeable soils grades varying between 0 and 2.0 percent are acceptable. On less permeable soils a minimum grade of 0.2 percent is desirable.

The distance the runoff must be carried by the crop row should be limited so that it does not accumulate in sufficient volume to break over the rows. The location of natural depressions which are converted to grassed waterways usually determines the length of the row.

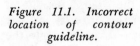

Figure 11.1. Incorrect location of contour guideline.

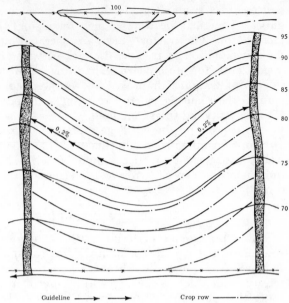

The most common mistake in locating contour guidelines is to place one near the middle of a slope, and plant all crop rows parallel to it. This *incorrect* way to locate guidelines is illustrated in Figure 11.1. Within a short distance on both sides of the guideline, the grade in the crop rows planted parallel to it will be such that the row drainage is toward the center of the field. The crop rows in most of the field thus drain water away from the waterways and concentrate it in the center of the field, resulting in severe gully erosion at this point.

The topography represented in Figure 11.1 is quite typical. Near the top, the land slope in the depressions is steeper than on the ridge between the depressions. Near the bottom the land slope in the depressions flattens out and is less than on the ridge. At some point near the center, the land slope on the ridge is approximately the same as in the depressions.

Two important considerations in the location of guidelines follow:

1. Where the slope is steeper in the waterways than down the ridge between them, the grade in the crop rows will increase as they are planted downslope parallel to a guideline and will decrease as they are planted up from a guideline. If a guideline is laid out to a minimum grade on this type of topography, it is necessary to plant crop rows down from it.

2. Where the slope is less in the waterways than down the ridge between them, the grade in the crop rows will decrease as they are planted downslope parallel to a guideline and will increase as they

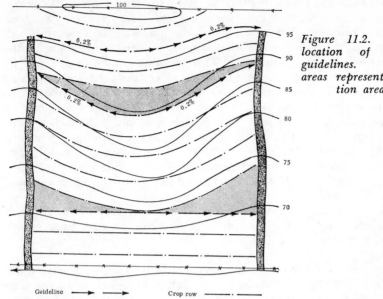

Figure 11.2. Correct location of contour guidelines. (Shaded areas represent correction areas.)

are planted up from a guideline. In this case, if crop rows are to be planted down from a guideline, it must be laid out to a maximum grade.

The *correct* way to locate guidelines is illustrated in Figure 11.2. The first guideline is located approximately one terrace spacing from the top of the slope. On this topography the guideline would be laid out to the minimum grade.

The distance down from this guideline that crop rows can be planted before the grade becomes excessive is estimated by using the following equation:

$$\text{Distance} = \frac{(\text{length of row})(\text{maximum grade} - \text{minimum grade})}{(\text{difference in land slope between waterway and ridge})}$$

The estimated distance down the slope is measured perpendicular to the guideline near the waterway and on the ridge. The difference in elevation between these points divided by the distance between them will give the overall grade in a crop row at that location. If this grade is greater than the maximum, the grade is rechecked from an upslope position, and if less than the maximum, it is rechecked from a downslope position.

If the land slope in the waterway is still greater than down the ridge, another guideline is laid out to a minimum grade. This will result in a correction or irregular-shaped area. The grade in the crop rows planted parallel to this guideline will increase until the land slope in the waterways becomes less than on the ridge. In Figure 11.2 this occurs at about the 80 contour line. Below this point the grade in the crop rows will decrease. At some distance down from the guideline the grade in the crop row will be at a minimum. This location can be determined by measuring down from the guideline a given distance and checking the grade in the crop row at this location as explained above. At the point where the grade in the crop row becomes minimal, another guideline must be staked out. This guideline should be at such a grade that crop rows parallel to it will drain water to the waterways. Normally this guideline would be staked out to a maximum grade, but in this example the lower field boundary was used as a guide and a crop row planted up from it. All crop rows have drainage to the waterway, and the area is easy to farm.

Good judgment and a knowledge of the topography is essential in locating guidelines that will result in proper row drainage and provide the most farmable layout.

Planting with Guidelines. In plowing with a moldboard or disk plow, a backfurrow should be thrown up on each guideline to mark the lines for future reference. If the correction areas are planted to a row crop, the rows should be planted parallel to whichever guideline will result in the rows draining in the direction of the waterways.

STRIP-CROPPING

In this practice, cultivated and close growing crops are planted in alternate strips across the slope. The runoff from the cultivated crop is retarded by the close growing crop, resulting in greater absorption of runoff and deposition of sediment. For this reason contour strip-cropping is more effective than contouring alone in reducing soil erosion. (See Ch. 4 for information relative to the effectiveness of contour strip-cropping in reducing soil erosion if information is not available for specific conditions in the area being considered.) This practice is most effective in areas of moderate rainfall, permeable soils, and short uniform slopes.

Buffer Strip-Cropping. If the correction areas in Figure 11.2 were seeded to grasses or legumes, they would form buffer strips between the strips of cultivated crops. The number and location of buffer strips are selected to give the desired protection from erosion.

Contour Strip-Cropping. The field is laid out much the same as for contour farming, except that the correction strips are made larger. The guidelines and strip boundaries should be so located that the row grade in all strips will be within permissible limits and will be toward the waterways. For convenience in carrying out the cropping system, the

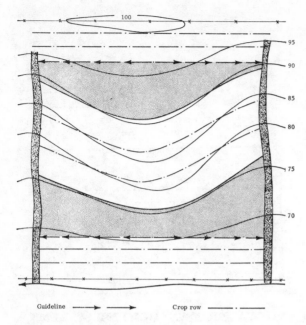

Figure 11.3. Contour strip-cropping.

Guideline ————→ ————→   Crop row ———— · ————

strips should be approximately equal in area (Fig. 11.3). This is the most effective method of strip-cropping.

Field Strip-Cropping. In field strip-cropping, strips of uniform width are laid out across the slope parallel to one guideline. This method is not as effective as others and should only be considered as a temporary control measure. The strips may also run parallel to the field boundary most nearly across the slope (Fig. 11.4). In most cases this is an even less effective method of control.

Strip Width. The recommended width of strips is given in Table 4.12.

MANAGEMENT OF RUNOFF WATER

Under certain conditions sheet erosion may be reduced to an acceptable level by a combination of soil and crop management, tillage methods, and contouring or strip-cropping. However some additional provisions must be made to prevent erosion by the concentrated runoff water. Grassed waterways may be constructed in depressions where runoff water accumulates, or a system of earth fills may be constructed across depressions to collect runoff from crop rows and discharge it through underground outlets. The earth fills may have both slopes flat enough to farm or they may be constructed with a steep backslope and only the frontslope farmed (11). The system of earth fills and underground outlets would effectively reduce the rate of runoff, sediment loss, and possibility of sediment pollution.

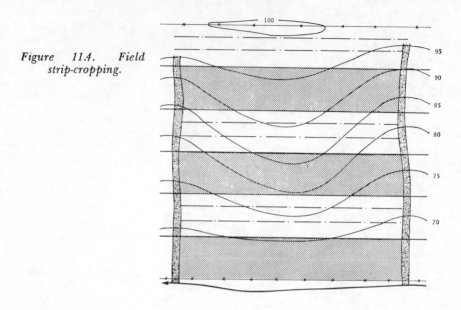

Figure 11.4. Field strip-cropping.

PLANNING SOIL AND WATER MANAGEMENT SYSTEMS

A well-planned soil and water management system must control erosion and conserve water. It must also provide for efficiency in farming operations, contribute to increased productivity, and provide an adequate income. In planning such a system, the following information will be helpful: type of soil and extent of erosion; a topographic map of the farm for planning the location of waterways, diversions, structures, roadways, and field arrangements; the cropping system planned for the farm; number and size of fields necessary for the farm enterprises; soil and crop management practices to be followed; tillage methods to be used; expected quality of maintenance and management; income potential of the farm; and available finances.

In developing soil and water management systems some of the methods to be considered for erosion control are permanent vegetative cover, crop rotations, mulch tillage, ridge planting, contouring, strip-cropping, broad-base terraces, steep-backslope terraces, flat-channel terraces, bench terraces, cross-slope channels, diversions, grassed waterways, underground outlets, water detention fills with underground outlets, and stabilization and flood retardation structures. The methods used should be adapted to the soil, topography, and climate of the area and should enable the landowner to operate efficiently the type of farming enterprise that will meet his needs. The procedure given in Chapter 4 may be helpful in determining which method or combination of methods will give adequate erosion control. In most situations, a combination of the above control measures will be required. However, effective control of raindrop erosion and erosion resulting from surface flow will reduce the need for the more expensive downstream control measures. Thus those measures which reduce raindrop erosion and sheet erosion should receive first consideration in an erosion-control program.

After the methods have been selected, certain requirements should be recognized in developing an efficient soil and water management system.

1. To facilitate the movement of livestock and machinery from the farmstead to the fields and from field to field with a minimum of travel and a minimum number of gates to open and close, convenient access to all fields should be provided.
2. To meet the changing needs of crop and livestock enterprises, flexibility in the number and size of fields is necessary.
3. To provide for efficient use of machinery, rows are made as long as possible, and short or point rows are kept to a minimum.
4. If terraces are to be used—
   a. The terrace cross section that will meet the need for soil and water conservation and permit efficient operation of the largest machinery to be used on the terraced land should be selected.
   b. The terraces or sections of terraces are made parallel, with long, gentle curves wherever practical.
   c. The type of outlet or outlets that best meet the needs for adequate erosion control, sediment pollution control, moisture conservation, and the needs of the farm business are selected.

      Grassed outlets are low in first cost compared to underground outlets, but they do take up space that has previously been available for grain crop production, and they often interfere with the operation of farm machinery. The use of chemicals for weed control on adjacent land is causing increased problems in grassed waterway maintenance.

      Underground outlets should be considered (1) on highly productive land devoted to grain crop production if the farmer has no need for the forage produced in a grassed waterway, (2) where an overfall structure would be eliminated by use of the underground outlet, (3) where a grassed waterway would require drainage, and (4) where a grassed waterway would be difficult to construct and maintain. Underground outlets reduce peak rates of runoff and sediment losses from the field.
   d. The best location should be selected for the outlet. If a grassed waterway is used, a location should be selected that will result in maximum farmability of the terraced area and a reasonable cost for outlet construction. The following factors deserve consideration:
      (1) The upper ends of the terraces should be at the end of the field most accessible to the farmstead.

(2) Crop rows should be kept as long as possible.

(3) Extreme variations in the slope of the land to be terraced into the outlet should be avoided. This will result in terraces that are less crooked and more uniformly spaced.

(4) Locations where the outlet will be higher than the land adjacent to it should be avoided. A cut must be made in constructing the terraces and if the outlet is on higher ground the terraces must be curved downslope to enter the outlet, resulting in an area that is difficult to farm.

(5) Locations where the outlet will be difficult or costly to construct and maintain are to be avoided.

(6) An outlet should never be located where it will be used as a roadway or as a livestock lane.

(7) A location should be selected where the terraces will best fit the topography of the area and can be made as parallel as possible.

e. Terraces, outlets, diversions, and other structures are located where they will offer the least possible interference to the farming operations.

5. If terraces are not used, a combination of cropping system and tillage method is selected that will reduce sheet erosion to an acceptable level. Also, in order to prevent serious erosion, provisions must be made to control the accumulation of runoff from the crop rows. This may be accomplished by a system of grassed waterways in the depressions to control the runoff from the crop rows, or a system of earth fills constructed across the depressions to collect the runoff from the crop rows and discharge it at a low rate through underground outlets.

6. Ponds and reservoirs should be located conveniently for the intended uses. The water management system is planned to provide an adequate source of water for these impoundments.

7. Most farmers will wish to have some areas on the farm suitable for wildlife cover or for recreation. On most farms there will be areas not too desirable or accessible for crop production that can be used for these purposes.

8. In order to prevent damage to adjoining property by the conservation practices utilized, the discharging point of surface water from the water management system should be the same as the one existing naturally before the system was established; and surface water should not be impounded on adjoining property or forced to assume another course in leaving the adjoining property. If it is not possible to plan a system following these principles, an agreement with the owner of the adjoining property should be reached before the water management system is constructed.

9. The system should be planned for maximum benefits at reasonable costs.

10. It is seldom possible to develop the best plan at the first attempt. Various alternative solutions should be analyzed before the final plan is adopted.

PROBLEMS

11.1. Prepare a scorecard on which you can rate the various tillage methods as to how well they meet the requirements for ideal tillage. List the methods across the top of the scorecard and the requirements for ideal tillage down the left side. Select a specific crop in a cropping system in your area and rate the various tillage methods on the scorecard, using values from 1 to 10. Assign the value of 10 if the method is considered ideal for that particular requirement.

11.2. Use transparent paper as an overlay for the map of the farm shown in Figure P11.1. Develop a soil and water management system for this farm, which is located in northwest Missouri. The soil is a Marshall silt loam which has a dark colored, well-aerated surface soil with good water-holding capacity and is quite productive. Erosion has reduced the depth of the surface soil to less than 8 inches on parts of the farm. The subsoil is a brown, silty clay loam extending to a depth of 6 to 8 feet. The owner is a good manager and has adequate finances; nevertheless, he is interested in maximum income from his enterprise.

    a. Develop a soil and water management system for this farm if terraces are not used to control erosion. Indicate the soil and crop management practices to be followed, the tillage methods to be used, and the method to be used to control the runoff water.

    b. Develop a soil and water management system for this farm if terraces and grassed waterways are used to control erosion. Plan the system so that the owner can produce row crops and graze cattle on practically the entire farm. Show the location of terraces, waterways, diversions, structures, ponds, fences, gates, and lanes.

    c. Analyze the plan shown in Figure P11.2 to determine how well the principles previously discussed were followed in developing this plan.

11.3. Erosion on the farm shown in Figure P11.3 has been moderate to severe as indicated by the drainage patterns. Use transparent paper as an overlay and develop plans for this farm with conditions as specified in (a) and (b). In each plan (1) indicate the soil and water conservation practices that will be used; (2) show the location of terraces, grassed waterways, underground outlets, diversions, structures, and ponds if they are used; and (3) show the location of fences, gates, and lanes.

    a. Plan a soil and water management system if the following conditions exist:

        1. The topsoil is quite shallow and the productivity level of the subsoil is low.

        2. The cropping system will consist of approximately two-thirds grain crops and one-third forage.

        3. Livestock will be kept to utilize the forage.

        4. Conventional tillage methods will be used. Four-row equipment is being used.

        5. Maintenance and management possibilities are fair to good.

        6. Finances are limited.

        7. Income potential of the farm is fair.

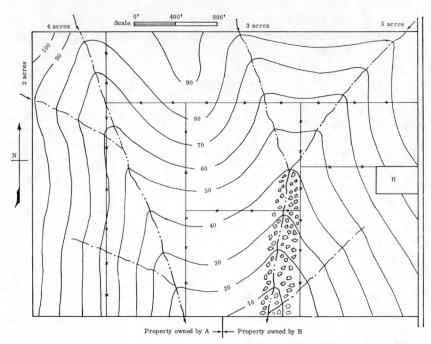

Figure P11.1. Topographic map for use in planning a soil and water management system.

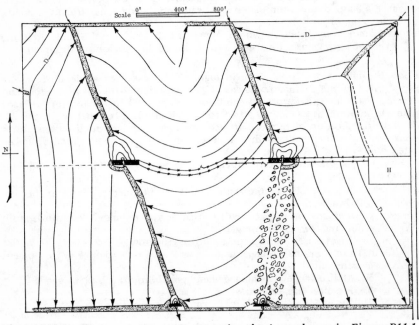

Figure P11.2. Water management system for the farm shown in Figure P11.1.

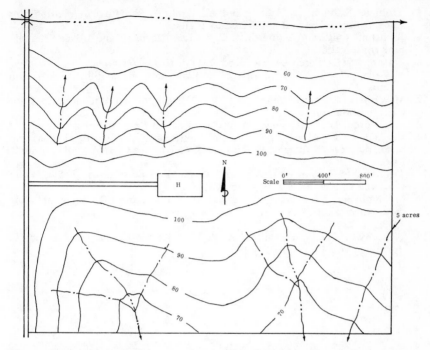

Figure P11.3. Topographic map for use in planning soil and water manage-ment systems.

b.   Plan a soil and water management system if the following conditions exist:
   1.   The topsoil is quite deep and the subsoil responds well to fertility treatment.
   2.   This will be a cash-grain operation with no livestock.
   3.   This proprietor farms a large acreage in addition to this and is interested in maximum efficiency in his operations. He uses six-row equipment.
   4.   He is presently using conventional tillage practices but is consider-ing some type of minimum tillage.
   5.   Maintenance and management possibilities are good.
   6.   Finances are adequate.
   7.   Income potential of the farm is good.

REFERENCES

Note: References 1 through 9 are taken from American Society of Agricultural Engineers Publication Proc–168, Tillage for greater crop production, 1967.
1.   Bertrand, A. R.   Effect of tillage on soil properties and water content.
2.   Buchele, W. F.   Mechanization of soil erosion control practices.
3.   Burnett, Earl, and Hauser, V. L.   Deep tillage and soil-plant-water rela-tionships.

4. Holt, R. F., Voorhees, W. B., and Allmaras, R. R. Nutrient relationships and fertilizer placement as affected by tillage.

5. Johnston, J. R., and Van Doren, C. E. Land forming and tillage for moisture conservation.

6. Larson, W. E. Potential and need for soil tillage research.

7. McCalla, T. M. Effect of tillage on plant growth as influenced by soil organisms.

8. Meyer, L. D., and Mannering, J. V. Tillage and land modifications for water erosion control.

9. Van Doren, D. M. Changes in seed environment due to tillage.

10. Burwell, R. E., Allmaras, R. R., and Sloneker, L. L. Structural alteration of soil surfaces by tillage and rainfall. *J. Soil Water Conserv.*, vol. 21, no. 2, 1966.

11. Jacobson, Paul. Soil erosion control practices in perspective. *J. Soil Water Conserv.*, vol. 24, no. 4, July–August 1969.

12. Moldenhauer, W. C., and Amemiya, M. Tillage practices for controlling cropland erosion. *J. Soil Water Conserv.*, vol. 24, no. 1, January–February 1969.

13. Wittmuss, H. D., and Swanson, N. P. Till planted corn reduces soil losses. *Agr. Eng.*, vol. 45, no. 4, May 1964.

12 ～ Planning for Urban Development

Changes in land use brought about by a growing population concentrated largely in expanding urban centers are causing serious erosion and sediment pollution problems.

THE expanding population has created a need for new houses, factories, shopping centers, schools, highways, water supply systems, sewage systems, and recreational areas. To meet these needs, the topography has been changed, vegetative cover has been destroyed, and areas are covered by buildings or paving. These changes have resulted in decreased infiltration, increased runoff, erosion, sediment pollution, and flooding (2). Often these changes have been made with no thought to the quality of the environment which results. Man-made facilities have been crowded into all available space, destroying the beauty of the countryside and polluting streams and lakes.

When erosion is not controlled during the urban development period, damages can be extreme. Eroded slopes require regrading and additional stabilization. Highways, foundations, and other improvements may be undermined by erosion. These and many other damages result in unnecessary delays in projects, expensive repairs, and an unattractive landscape.

Sediment, the product of this erosion, causes additional problems. When consideration is not given to reducing erosion during urban development, sediment losses up to 100,000 tons per square mile have occurred. This sediment fills in and pollutes streams and reservoirs, reducing their capacity and destroying their scenic and recreational value. Storm drainage systems are filled in, and flooding is increased. The removal of this sediment is difficult and expensive.

PLANNING

Developers of urban areas must include practices for erosion control, safe disposal of runoff, and preservation of the natural beauty of

263

the landscape in their overall planning. Planning and implementing these practices will result in reduced maintenance and corrective costs and add to the well-being and desirability of the community. It will also result in happier neighborhoods downstream, which often bear the brunt of the effects of indiscriminate development.

A community or region in the development of a comprehensive plan for expansion should consider the needs of the homeowner, agriculture, business, industry, commerce, recreation, and aesthetics. The necessary regulations, ordinances, and codes should be adopted so that (1) erosion and sedimentation are effectively controlled during development, (2) adequate drainage and control of runoff can be provided, (3) the land is used according to its capability, (4) construction meets acceptable standards, and (5) aesthetic values are preserved.

TOPOGRAPHY

Maximum consideration should be given to the natural topography in planning developments. Steep areas which would erode seriously if disturbed should be retained in their natural condition and used as open scenic areas, or if suitable, for parks and recreation. Areas subject to flooding or to a high water table should be considered as storage basins to reduce flood runoff, as open green space, or as sites for recreational facilities. Roads, streets, and drainage facilities should be planned and located so that construction and maintenance costs and potential for erosion and runoff are minimized. If good reservoir sites are available, they should be considered as a possible means of reducing runoff peaks and as recreation and wildlife habitat areas.

SOILS AND GEOLOGY

In making choices for use of the land, planners and developers need all the information they can obtain about the soils and geology of the area and its limitations for various uses. Detailed soil and geologic maps, interpretive reports, and on-site investigations by a soil scientist and geologist provide information useful in evaluating the physical environment of an area and in selecting the proper land use. Specifically, they are helpful in determining (1) the load-bearing capability of the soil and bedrock, and its suitability as building sites for homes, schools, industry, or roads; (2) the shrink-swell potential of the soil under varying moisture conditions and how this may affect foundation stability; (3) the permeability of the soil and its adequacy for sewage disposal; (4) the depth and physical characteristics of the soil as they affect the comparative grading and excavation costs for roads and buildings at different locations, and the availability of rock, gravel, and sand which can be used for construction; (5) the drainage characteristics of the soil and the need for drainage facilities or for extra precautions in waterproofing basements; (6) the layering and presence of undesirable characteristics in soils which would make them susceptible to landslides; (7) the possibility of establishing and maintaining vegetation on the soil, and the treatments needed to improve the possibilities for a vegetative cover; (8) the available water-holding capacity of the soil and the need for irrigation to support a good vegetative cover; (9) the presence of salts, alkalinity, or acidity that would inhibit plant growth or affect construction materials; (10) the infiltration rate of the soil and its effect on rates of runoff and cost of storm drainage facilities; and (11) the erodibility of the soil and practices needed for control.

WATER MANAGEMENT

Water management problems are intensified by urban development. Natural cover is often destroyed during construction and replaced by buildings, streets, and parking areas. These changes reduce infiltration and increase runoff. The increased runoff results in higher water levels and flooding of land not previously flooded. Urban encroachment onto the floodplain necessitates building of levees and channels to confine the flow. The higher rate of runoff resulting from reduced infiltration plus the loss of overbank flood storage due to the levees and channels causes flood peaks downstream far in excess of those previously experienced.

Excessive erosion during development results in sedimentation of channels and reservoirs, thus causing even greater flood damages.

VEGETATION

Plans should be developed to retain and protect the natural vegetation whenever feasible and to establish vegetation on areas subject to erosion. A properly planned and executed maintenance program is necessary if a satisfactory vegetative cover is to be maintained.

DEVELOPMENT

After a comprehensive plan has been completed, there are a number of principles and practices that will prove effective in reducing the soil and water problems during the various stages of development.

SEQUENCE OF OPERATIONS

In too many instances, the first construction operation is to clear the entire area, exposing large expanses of bare soil to the erosive power of the rain and wind. This is sometimes done even though construction may not be completed on parts of the area for several years. The construction plan should keep the areas of exposed bare soil to a minimum for the shortest time. This requires clearing and grading in segments only as needed for construction. For example, in a residential development, only the street right-of-way is initially cleared and other areas are not disturbed until the storm drainage system has been installed. Maximum effort should be made to save existing vegetation insofar as possible.

STABILIZATION OF CRITICAL AREAS

Mulches, fibrous netting, quick-growing grasses, small grains, shrubs, or sodding should be used to stabilize areas subject to erosion as soon as possible. Topsoil may be stockpiled and spread back over critical areas to assist in establishing a vegetative cover. Consideration should be given to the soil, climate, steepness of slope, orientation of slope, shade, and traffic in selecting the type of permanent vegetation. With the wide variety of trees, shrubs, vines, grasses, and other types of cover available, it is usually possible to establish a vegetative cover of some type on almost any site.

BORROW AND SPOIL DISPOSAL AREAS

These areas should be selected carefully, giving consideration to their effect on runoff, erosion, and aesthetics, as well as convenience for construction. They should be graded to blend in with the surroundings and a good vegetative cover established as soon as possible after construction is complete.

DIVERSIONS, TERRACES, AND WATERWAYS

Diversions and terraces can be used to divert water from areas that would be subject to severe erosion or landslides and to intercept and divert surface runoff before it accumulates in sufficient volume to cause serious erosion. Grassed waterways should be used throughout the area to dispose of surface runoff.

BENCHES

On steep topography, it is common practice to form level benches on the slope by cutting and filling, resulting in very steep slopes between the level benches on which buildings are located. This should never be attempted unless a thorough analysis is made of the soils and geology of the area to determine the possibility of causing landslides or soil creep following construction.

SEDIMENT BASINS

It is seldom possible to avoid all erosion while an area is being developed. For this reason a sediment trap or basin should be constructed downstream from the construction area to remove the sediment from the runoff water. These structures usually consist of an earth fill with a pipe and emergency spillway. The sediment space provided should be based on the expected yield from the area during the development period. If sufficient space cannot be provided, provisions should be made for periodic cleanout.

SUBSURFACE DRAINAGE

Subsurface drains may be needed to eliminate wet areas during construction, to intercept underground runoff which would contribute to landslides and soil creep, to eliminate wet basements, to permit the growth of desirable vegetation, and to generally improve livability of the area.

WATER STORAGE STRUCTURES

If storm runoff can be stored and released at a slower rate, the capacity of the system to carry it can be reduced. Storage may be provided in natural areas such as floodplains, lakes, and marshes, and also in flood retardation structures if adequate sites are available. The permanent pool in the flood retardation structure also provides recreation and a wildlife habitat.

Impounding runoff water on individual lots or segments of land effectively retards surface runoff in areas having permeable soils. This water soaks into the ground and not only reduces surface runoff but provides additional moisture for the growth of vegetation. This method is particularly effective in areas subjected to widely distributed intense rains.

OPEN CHANNELS

The carrying capacity of natural channels can be improved by straightening, increasing the size, and lining with stone or concrete, and by removing brush, trash, and other obstructions from the channel. In most cases, the appearance can be improved along with increasing the carrying capacity.

EMERGENCY SPILLWAYS

Seldom is it economically feasible to carry the expected maximum runoff in the primary or constructed storm drainage systems. Provisions *must* be made to handle the runoff that exceeds the capacity of the primary system, without excessive damage to property or loss of life.

In structures used to control runoff from agricultural land, the primary system is the principal spillway, and the excess runoff is carried in an emergency spillway. In like manner, emergency spillways must be provided to carry the excess runoff from urban areas. These spillways should be located so that occasional flow will not create a hazard or cause excessive damage. They should be maintained in such condition that the channel will remain unrestricted and protected from erosion.

DESIGN CRITERIA FOR SOIL AND WATER MANAGEMENT SYSTEMS

Information needed for the design of various soil and water management systems has been presented in previous chapters. This same information, with modifications where needed, can be used when developing urban areas.

REFERENCES

1. American Society of Planning Officials. Soils and land-use planning. Papers presented at the ASPO planning conference, Philadelphia, April 17–21, 1966.
2. Barnes, R. C., Jr. Erosion control practices adapted for urban use and pollution abatement. Paper 69–223, presented at the American Society of Agricultural Engineers' meeting, Chicago, December 1969.
3. Hanke, B. R. Planning, developing, and managing new urban areas. In *Soil, water, and suburbia.* USGPO, Washington, March 1968.
4. Ifft, T. H. Solutions to urban-fringe erosion-sedimentation problems. *Agr. Eng. J.,* vol. 51, no. 12, December 1970.
5. McHarg, Ian L. *Design with nature.* Garden City, N.Y.: Natural History Press, 1969.
6. Phillips, R. L. Solutions to urban-fringe hydrology and drainage problems. *Agr. Eng. J.,* vol. 51, no. 12, December 1970.
7. USDA. Soil, water, and suburbia. A report of the proceedings of a conference sponsored by the USDA and the U.S. Department of Housing and Urban Development, Washington, D.C., June 15 and 16, 1967.

13 ～ Measuring Distances, Areas, and Volumes

In planning and laying out soil and water conservation systems it is necessary to measure distances, the area of watersheds and fields, and the volume of earth in excavations and embankments.

THE METHOD and equipment used in measuring distances depends on (1) the accuracy required in the work for which the measurements are made, (2) the equipment available, (3) the size of the object or area to be measured, and (4) the condition of the terrain. The necessary precision required for the work should be determined and the method and equipment used which will most conveniently secure the measurements needed.

EQUIPMENT

Measuring wheel. This wheel is constructed so that one revolution requires a movement of 10 feet or in some cases, 1 rod. The revolutions are counted as the wheel is pushed along the course to be measured. An odometer may be attached to a wheel to register the number of revolutions. The circumference of the wheel being known, the distance passed over can then be computed. An error of 1/200 may be expected.

Stadia. Transits and levels may be equipped with stadia hairs, one above and one below the level cross hair, spaced so that the distance the rod is held from the instrument may be determined. Most stadia hairs are so spaced that when an interval of 1 foot on the rod is spanned by the upper and lower cross hairs, the rod is approximately 100 feet away; the cross hairs span 2.75 feet on the rod when it is 275 feet away. On externally focusing instruments a stadia constant, in most cases approximately 1 foot, must be added to the stadia measurement to secure correct distances. For example, if 1 foot is spanned on the rod the true distance would be 101 feet. If 2.75 feet were spanned the true distance would be 276 feet. On internally focusing instruments a stadia constant is usually not added. With stadia an error of 1/500 to 1/1,000 may be expected depending upon the quality of the instrument and care in its use.

Electronic devices. Recently developed electronic devices are valuable for the precise measurement of long distances. Basically they measure the time for an induced wave to reach a reflector and return to the sender. They are not widely used in soil and water conservation work, since it is seldom necessary to measure long distances with extreme accuracy.

Chain. Much of the early surveying in this country was done with the surveyor's (Gunter's) chain. This chain was 66 feet long and was made up of one hundred links, each 0.66 foot long. It was convenient for land surveying because the following relationships exist:

$$1 \text{ chain} = 4 \text{ rods}$$
$$80 \text{ chains} = 1 \text{ mile}$$
$$10 \text{ square chains} = 1 \text{ acre}$$

Later the engineer's chain was developed. It was similar to the Gunter's chain, except that it consisted of one hundred links, each 1 foot long. Both of these chains were subject to wear in the many link connections and required frequent adjustment.

Steel tape. Improvements in the quality of steel made possible the manufacture of thin steel ribbons called tapes. These may be graduated in feet and tenths and hundredths of a foot; in feet, inches, and fractions of an inch; in meters, decimeters, and centimeters; or in chains. Surveying equipment graduated in feet and tenths and hundredths of a foot is most commonly used in the United States. Tapes come in various lengths, but the 100-foot tape is most commonly used. The end foot is divided into tenths and sometimes hundredths of a foot, numbered from zero toward the 1-foot mark and from the 99-foot toward the 100-foot mark. Some have an extra foot beyond the zero end divided into tenths and hundredths. Care must be exercised in locating the zero and 100-foot mark on the particular tape being used. Many place the zero point at the extreme end of the loop to which the end ring is attached, instead of using a graduation on the steel ribbon.

Care must be taken that the steel tape is never pulled taut if it has a coil or a kink in it, as it is very easily broken. After each use the steel tape should be wiped off with a dry rag and a light film of oil applied before it is stored.

Pins or arrows, eleven to a set, are used to stick in the ground to mark definite points along the tape in measuring distances. The pins are also used to keep a record of the number of tape lengths measured. They should be kept on the ring provided, except when they are being used to mark a point.

Range poles are made of metal, wood, or fiberglass. They are of octagonal or circular cross section, usually 8 feet long, with alternate bands of red and white, each being 1 foot long. They are used to indicate the location of points or the direction of the line to be measured.

METHODS

Stepping. When the accuracy required by the work is not too great, distances may be measured by stepping, which consists of counting the number of steps in the distance being measured and multiplying this number by the length of steps. The length of step is determined by counting the steps in a known distance while walking in a normal manner. The length of step varies with the speed of walking, the slope of the land, and the condition of the surface; these factors should be taken into consideration. Under favorable conditions an error of 1/50 may be expected.

Taping. Measurement by taping is the most widely used method of measuring distances when accuracies in the magnitude of 1/1,000 to 1/25,000 are required.

Ground surfaces. In measuring the distance between points, the head tapeman uncoils the tape, takes the zero end and ten pins and drags the tape toward the distant point. One pin is kept by the rear tapeman even though it may not be needed to mark the starting point. When the 100-foot end of the tape is near the starting point, the rear tapeman calls "tape" and picks up the tape. The head tapeman stands to one side of the tape and holds a pin against it, so that the rear tapeman can align the pin on the point to which they are measuring. When exactly in line the head tapeman places the pin on line near the end of the tape. He then straightens the tape along the ground and exerts sufficient pull so that the tape will lie taut. When the rear tapeman has the 100-foot mark exactly on his pin, he calls "stick"; the head tapeman then removes his pin from the temporary alignment position and places it exactly on the zero mark and calls "stuck." The rear tapeman then drops his end of the tape, pulls the pin, and proceeds to the pin just set by the head tapeman, where the process is repeated.

It is noted that when the head tapeman places a pin at the end of a full tape length, the rear tapeman pulls a pin from the ground. At any time the number of pins in the rear tapeman's possession represents the number of tape lengths measured. There is always one pin in the ground which marks the end of the distance measured but which is not counted. When 1,000 feet have been measured, the rear tapeman passes his ten pins forward to the head tapeman and makes a record of the distance measured.

At the end of the course there is usually a fractional part of a tape length to be measured. The head tapeman stops at the point marking the end of the course and holds the tape about halfway between the zero and the 1-foot mark. The rear tapeman then moves the tape to the nearest foot mark and calls out the number he is holding on the pin (for example, "48"). The head tapeman tightens the tape and reads the fraction of a foot at the end point (0.8, for example). The distance would be 48 minus 0.8, or 47.2 feet. If the tape had an extra foot beyond the zero end graduated into tenths, then the rear tapeman would have held the 47-foot mark and the head tapeman would have read 0.2 at the end of the course. In this case the 0.2 foot is added to the 47 feet. At the end of a course the two pins marking the last fractional tape length are left in the ground until the pins held by the rear tapeman have been counted and the distance recorded. For example, in the above illustration if the rear tapeman had eight pins in his possession the distance would have been 847.2 feet.

Horizontal distances. All land surveys are based on horizontal rather than surface measurements. To obtain true land area or to survey a piece of property it is necessary to measure distances horizontally.

To measure the horizontal distance on sloping ground the tape must be held level and a plumb bob used to mark the point on the ground directly below the point on the tape. It is important that the tape be kept horizontal if accurate work is to be done. A third member of the party is sometimes used to determine when the tape is held level by aligning the midsection of the tape with horizontal lines of buildings or by observing a small level attached to the tape at the center point.

If the slope is so steep that the man on the lower end of the 100-foot tape cannot hold it level, it is necessary to "break tape." The head tapeman pulls the tape out the full 100-foot length and then comes back to a point at which the tape can be held level. When measuring down-slope, the head tapeman uses a plumb bob to locate the point on the ground. The point is marked with a pin, and the rear tapeman comes forward and holds the same fractional point on the tape that was held by the head tapeman and the head tapeman moves forward another section. This method is continued until a full 100-foot length is measured. In measuring upslope, the rear tapeman must use the plumb bob. In "breaking tape," care must be taken to insure that the pins used to mark the fractional parts of a 100-foot length are not included in the number of pins held by the rear tapeman. These pins should be returned to the head tapeman. In some cases a separate set of smaller pins is used to mark the fractional points.

If distances are measured on the ground surface and the land is sloping, a correction must be subtracted from the surface distance to secure the true horizontal distance. The correction to be deducted for each 100 feet on different slopes is given by the formula $S^2/200$, in which S is the land slope in feet per 100 feet (slope percentage). This formula is not exact, but the error introduced by its use on slopes under 30 percent is less than 0.1 percent.

Sources of error. Following are some of the more common sources of error in measuring horizontal distances with a steel tape:

1. Length of tape. A tape must be of known and definite length. A 100-foot tape should be 100 feet long when lying on a flat surface, at a temperature of 68° F, and under a 10-pound pull.
2. Sag in the tape. As a tape is lifted from a flat surface the sag in the tape will shorten the distance between the end points. The shortening due to sag for a 100-foot tape with a 10-pound pull exerted on the ends will be approximately 0.06 foot for lightweight tapes ¼ inch wide, and 0.10 foot for heavier weight tapes 5/16 inch wide. In a 50-foot length the shortening due to sag would be approximately 0.01 foot for lightweight tapes and 0.02 foot for heavier weight tapes. If an error of less than 1/1,500 is required, the sag should be computed and the distance corrected.
3. Tape not in alignment. This error can be avoided by being careful to measure in a straight line between points and by holding the ends of the tape at equal elevations when measuring horizontal distances on sloping ground.
4. Variation from standard temperature. Most tapes are 100 feet long at a temperature of 68° F, and will expand or contract about 0.01 foot for each 15° F change in temperature. If an error of less than 1/5,000 is required, this correction should be made.
5. Personal errors such as not using the proper zero and 100-foot marks at the ends of the tape, careless plumbing, improper tension on the tape, improper setting of the pins, and many others can be avoided by careful work, or by measuring a distance more than once.

AREAS

If maps or aerial photographs are available for the area being planned, the areas of the fields or watersheds may be determined by taking measurements on the map or photograph or by using a planimeter. A planimeter mechanically integrates an area and records the area on dials as a tracing point is moved around the outline of the area to be measured.

If maps or photographs are not available, it is necessary to take measurements in the field to determine areas. Areas are usually expressed in acres (43,560 sq ft).

Rectangle. The area of a rectangle is the product of the length and the width.

Triangle. The area of a triangle (Fig. 13.1) may be determined in a number of ways. Two methods commonly used are—

1. If the length of the base and the height can be determined,

$$\text{Area} = \frac{\text{base } (b) \times \text{height } (h)}{2}$$

2. If the length of the three sides can be determined,

$$\text{Area} = [S(S-a)\ (S-b)\ (S-c)]^{1/2}$$

$$\text{where }\ S = \frac{a+b+c}{2}$$

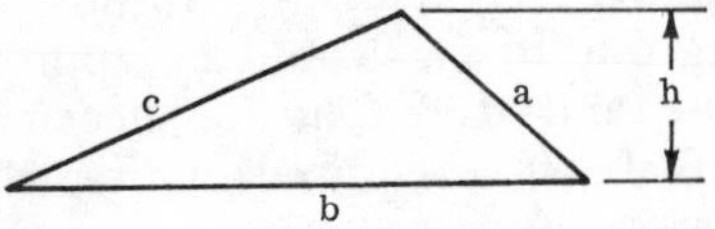

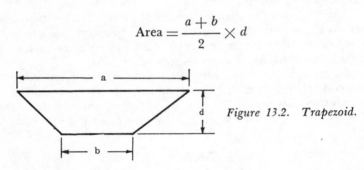

Figure 13.1. Triangle.

Trapezoid. A trapezoid has four sides, two of which are parallel (Fig. 13.2). If the length of the parallel sides and the distance between them can be determined,

$$\text{Area} = \frac{a+b}{2} \times d$$

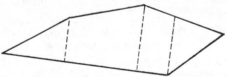

Figure 13.2. Trapezoid.

Many-sided figure.

1. The area may be divided into triangles, trapezoids, or rectangles and the areas of these determined and totaled. (See Fig. 13.3.)

Figure 13.3. Many-sided figure divided into triangles and trapezoids.

2. The length of all sides and the size of all angles may be determined and a map of the area prepared. The area can then be determined by using a planimeter.

Figure with a curved boundary. Establish a straight line near the curved boundary, line *A-B* in Figure 13.4. Measure perpendicular distances from line *A-B* to the curved boundary. Space these measurements so that the curved segment between the measurements can be considered a straight line without introducing a significant error. The areas can then be computed as trapezoids or triangles.

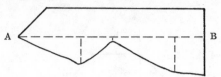

Figure 13.4.   Figure with a curved boundary.

If the area is bounded by a relatively smooth curve, the measurements perpendicular to a line *A-B* may be taken at equal distances as indicated in Figure 13.5. The area bounded by the curve is now divided into a series of trapezoids of equal altitudes and their combined area is

$$\text{Area} = d(h_o/2 + \Sigma h + h_n/2)$$

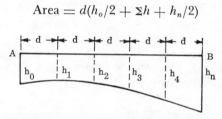

Figure 13.5.   Figure with a smoothly curved boundary.

where $d =$ distance between perpendicular measurements
$\Sigma h =$ sum of all perpendicular measurements except the two ends
h_o and $h_n =$ the end measurements

Cross section of embankments or excavations. In constructing embankments or making excavations it is necessary to obtain the cross-sectional area at specific points so that the volume of earth to be moved can be determined.

For a three-level section, if the depth of cut or height of fill is given at the center of the excavation or embankment, and at the points where the side slopes intersect the original ground, the area is computed by means of the formula given in Figure 13.6.

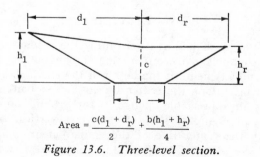

$$\text{Area} = \frac{c(d_1 + d_r)}{2} + \frac{b(h_1 + h_r)}{4}$$

Figure 13.6.   Three-level section.

For a five-level section, if the depth of cut or height of fill is given at the points indicated in Figure 13.7, compute the area by using the given formula.

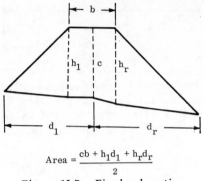

$$\text{Area} = \frac{cb + h_1 d_1 + h_r d_r}{2}$$

Figure 13.7.   Five-level section.

For an irregular section, if the depth of cut or height of fill is given at random points as in Figure 13.8, compute the area by dividing into triangles and trapezoids or using a planimeter.

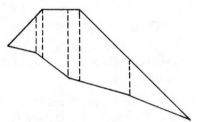

Figure 13.8.   Irregular section.

VOLUMES

It is often necessary to determine the volume of earth moved in the construction of soil and water conservation practices. The cubic yard (27 cu ft) is the standard unit for earth work.

CROSS-SECTION METHOD

This method is adapted to determining the volume of earth moved in excavating channels or in constructing embankments. Determine the height of the embankment, or the depth of the channel, above or below the ground surface along the center line of the embankment at points where there is an appreciable change in the ground slope. See h_1, h_2, h_3, h_4, and h_5 in Figure 13.9(a). Determine a vertical cross section at right angles to the center line of the embankment, or excavation, at these points. See Figure 13.9(b). The distance between these sec-

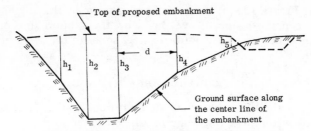

(a) Points at which vertical cross sections are taken to
determine the volume of earth in an embankment. Heights
of the embankment, along the center line of the embankment,
are designated h_1, h_2, h_3, h_4, h_5.

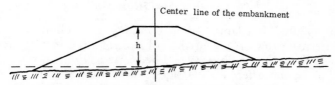

(b) A vertical cross section taken at right angles to the
center line of the embankment

*Figure 13.9.   Cross-section method of determining the volume of earth in an
embankment.*

tions should not exceed 100 feet when the land slope is less than 4
percent. This distance should be reduced to 50 feet on land slopes of
4 to 8 percent and to 25 feet on land slopes over 8 percent. Measurements should be taken so that the areas of these sections can be computed or measured with a planimeter.

The volume of earth between two vertical cross sections is equal to
the average area of the end sections multiplied by the horizontal distance
between them. For example, in Figure 13.9 the volume in the embankment between sections h_3 and h_4 would be equal to d(area at h_3 + area
at h_4)/2.

An approximate method known as the *average-height-of-section*
method is sometimes used to obtain the volume of earth in embankments. It is adapted to locations having no appreciable change in slope
along the center line of the embankment. The height of the embankment above the ground is determined at equal intervals, usually 10 feet,
along the center line. The area at each point is determined and multiplied by the distance between points (10) to determine the volume in
each section.

Another approximation that is sometimes made is to assume that
the land slope perpendicular to the center line is level, in which case
the base of the embankment is assumed to be parallel to the top of the
embankment, which gives a trapezoidal cross section. Since the height
and top width are known, the base width and the area of the trapezoid
can be computed. If the land perpendicular to the center line is slop-

ing, as in Figure 13.9(b) there may be an error in this method since the area between the assumed level line and the ground line on the right side of the center line may not be exactly equal to the corresponding area to the left.

CONTOUR-AREA METHOD

This method is adapted to determining from a contour map the volume of water in a reservoir, or in detention storage in a reservoir. The volume is obtained by measuring the area enclosed by each contour and multiplying the average area of adjacent contours by the vertical distance between contours.

GRID METHOD

This method is adapted to determining the volume of earth excavated from borrow areas and the volume required in fills over larger areas. The area to be excavated or filled is staked out in squares of 10, 20, 50, 100, or more feet on the side. The choice of grid size depends upon the size of the project and the accuracy desired. The elevation at each grid corner is determined prior to construction. After the construction is complete, determine the height of cut or fill at the four corners of each square.

In computing volume where all four corners of a square are either cut or fill, the average height of cut or fill at the four corners of each square is first determined. Then this average height is multiplied by the area of the square to determine the volume. The total volume is found by adding the volumes of the individual grids.

In computing volumes where both cuts and fills are to be made on the grid corners, the so-called four-point method is commonly used. The sum of the heights of cut at the grid corners c is determined. Then the sum of the heights of fill at the grid corners f is determined.

The cubic yards of cut can be determined by the equation

$$ V_c = \frac{A}{108} \left(\frac{c^2}{f+c} \right) $$

where A = area of the grid square in square feet.

Similarly the cubic yards of fill can be determined by the equation

$$ V_i = \frac{A}{108} \left(\frac{f^2}{c+f} \right) $$

PROBLEMS

13.1. What is the advantage of having measuring equipment graduated in feet, tenths, and hundredths of a foot instead of in feet, inches, and fractions of an inch?

13.2. In "breaking tape" why doesn't the head tapeman pull the tape forward to a point at which he can hold the tape level rather than pulling it out a full 100 feet and then coming back to that point?

13.3. Which of the errors in taping, if not corrected, will result in the distance recorded being greater than the true distance?

13.4. In checking his length of step a man found that he took thirty-seven steps in a 100-foot distance.
    a. What is his length of step?
    b. In measuring a given distance, this man took two hundred steps. Compute the distance.
    c. The correct distance in part (b) is 550 feet. Compute his percent error in stepping the distance.

13.5. In measuring a distance with a level with an internally focusing telescope, the rod reading at the upper stadia hair was 10.74 feet and at the lower stadia hair, 6.21 feet. What was the distance from the level to the rod?

13.6. The northwest quarter of the southeast quarter of section 10 has a uniform land slope of 10 percent in the north-south direction. The land is level in the east-west direction. How many acres of land are in this plot? How many acres of surface area are in this plot? Note: Refer to Figure 15.3 if you are not familiar with the subdivision of a section of land.

13.7. A rectangular field is measured by stepping and found to be 1,560 feet long and 660 feet wide. The man who made the measurements is able to step a distance within 2 percent of the correct distance. Based on his measurements how many acres are there in this field? This would be accurate to the nearest (0.01, 0.1, 0.5, 1.0) acre.

13.8 The area shown in Figure P13.1 is cut into fields 1, 2, and 3 by the two ditches. Use a scale to take the necessary measurements and compute the acres in each of the three fields.

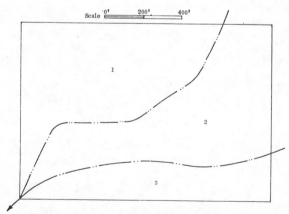

Figure P13.1. Map for use in Problem 13.8.

13.9. A drainage ditch constructed on nearly level land is assumed to have a trapezoidal cross section. The ditch has a 6-foot bottom width and eight-to-one side slopes. (An eight-to-one side slope means that for each 8 feet measured on the horizontal, the vertical distance is 1 foot.) At one point the ditch is 2.6 feet deep. A second point 100 feet from the first ditch is 3.2 feet deep. How many cubic yards will be removed in constructing this 100-foot section of the ditch?

13.10.   The embankment shown in Figure 13.9 has a 12-foot top width and three-to-one side slopes. Compute the cubic yards of earth in that section of the embankment bounded by h_3 and h_4, if $h_3 = 20$ feet, $h_4 = 8$ feet, and $d = 15$ feet—

 a. If the land slope perpendicular to the center line is level.

 b. If the land slope perpendicular to the center line is 4 percent.

13.11.   This sketch represents a part of a field that has been staked for land grading. The grid squares are 100 feet on each side. The depth of cut or fill is designated at each corner. Compute the volume of earth to be cut and the volume required to make the fill.

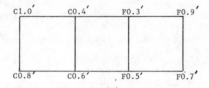

14 ~ Levels and Leveling

*The level is used to determine the vertical
distance or difference in elevation
between points. Its principal parts are a
level vial, often called a bubble tube,
and a telescope.*

THE LEVEL VIAL is used to level the instrument. It consists of a
glass tube, the top of which is shaped to conform to the arc of a circle.
The tube is almost filled with a sensitive, nonfreezing liquid; the small
amount of air forms a bubble at the high point on the curve. A line
tangent to the center of the vial when the bubble is in the center position is a level line.

The telescope consists of a tube with an eyepiece at one end, an
objective lens at the other, and a suitable arrangement of lenses and
cross-hair mountings between. The line of sight is an imaginary line
extending from the eyepiece through the intersection of the cross hairs.
The parts are so mounted that the line of sight and the bubble tube can
be made parallel. With the level properly adjusted the line of sight
through the telescope is level when the bubble is in the center of the
bubble tube; all points on this line have the same elevation. Some
levels are also equipped with a horizontal circle used in measuring
horizontal angles. To facilitate their use, levels, with the exception
of hand levels, are mounted on tripods.

TYPES OF LEVELS

A wide variety of levels is available, the following being most commonly used in soil and water conservation work:

The hand level (Fig. 14.1) is a low-precision instrument that is held
in the hand and is useful in making preliminary surveys and in rough
checks on work in progress. An error of approximately 0.1 foot in 30
feet can be expected.

The Abney hand level is equipped with a clinometer for measuring
the slope of the land.

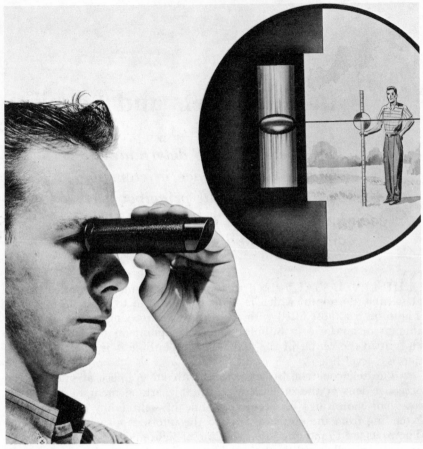

Figure 14.1. Hand level. (Courtesy Realist, Inc.)

Farm or contractor's levels are low-cost instruments for those jobs that do not require a high degree of accuracy. They are light and easy to handle, the telescope usually being about 10 inches long. They are of two general types—the dumpy and the turret. The dumpy level has the telescope and bubble tube rigidly attached to the frame of the level. The turret level has the telescope and bubble tube mounted in an invertible turret, thus facilitating the checking and adjustment of the level (Fig. 14.2).

Engineer's level is the term commonly applied to those levels sufficiently accurate for the layout of engineering practices. There is a wide range in the accuracy, quality, and cost of so-called "engineer's levels." There are two types—the wye and the dumpy. The wye level (Fig. 14.3) has the telescope and bubble tube combination mounted in two Y frames or yokes. This unit is held in place by clips that can be loosened for

*Figure 14.2.   Turret-type farm level.   Actual length—10 inches.   (Courtesy Bos-
trom-Brady Manufacturing Co.)*

*Figure 14.3.   Wye engineer's level.   Actual length—15 inches.   (Courtesy Tele-
dyne Gurley, Troy, N.Y. 12181)*

*Figure 14.4.   Dumpy engineer's level.   Actual length—15 inches.   (Courtesy Tele-
dyne Gurley, Troy, N.Y. 12181)*

removal of the unit, thus facilitating checking and adjustment of the
level.

The dumpy level (Fig. 14.4) has the telescope and bubble tube
rigidly attached to the frame. A dumpy level equipped with a horizontal
circle is shown in Figure 14.5.

Imported levels (Fig. 14.6) are usually very compact and lightweight.
Most are of the dumpy type and are equipped with a coincident bubble
that is viewed through the telescope eyepiece or through a separate eyepiece.

Automatic levels (Fig. 14.7) incorporate a self-leveling feature. The
instrument is set up approximately level by use of a circular spirit level,
and the line of sight kept exactly level by means of a pendulum and a
system of prisms and mirrors.

SELECTING A LEVEL

SENSITIVITY OF THE LEVEL VIAL

The sensitivity of the level vial is determined by the radius of
curvature to which the top of the vial is ground. The larger the radius,
the more sensitive the bubble. A highly sensitive bubble is necessary

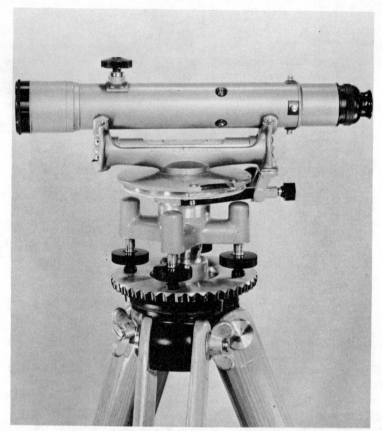

Figure 14.5. Dumpy level equipped with a horizontal circle. Actual length— 13 inches. (Courtesy Teledyne Gurley, Troy, N.Y. 12181)

for precise work, but may be considered a handicap on rough work because of the greater difficulty in centering the bubble. The more sensitive level vials are longer and ground to a high degree of precision. Uniformly spaced graduations are etched on the bubble tube to show the position of the bubble. These divisions are usually 2mm apart; however, a spacing of 0.1 inch has also been used.

The sensitivity of a level vial is defined as the vertical angle of rotation of the vial that moves the bubble one division as marked on the vial. It may also be expressed as the angle between the line of sight and a level line when the bubble is moved one division (Fig. 14.8).

The sensitivity of the bubble determines the accuracy. The error that can be expected from a level equipped with a 9-minute bubble if the bubble is off-center one division is shown in Figure 14.9. Note that the error increases as the distance from the level increases. The sensitivity of the level vial and the care the instrument man takes in centering the

Figure 14.6 Compact dumpy level. Actual length—8 inches. (Courtesy Kern Instruments, Inc.)

bubble at each reading are extremely important. If it is not centered for each reading, the amount of error involved will be in proportion to the sensitivity of the bubble, the amount it is off-center, and the distance from the level to the rod. By using extreme care an instrument man may be expected to keep the bubble within one-fourth of one division on the level vial. The errors expected with level vials of varying sensitivity are given in Table 14.1.

A number of levels are being equipped with a coincident bubble,

Figure 14.7 Self-leveling automatic level. Actual length—8 inches. (Courtesy Eugene Dietzgen Co.)

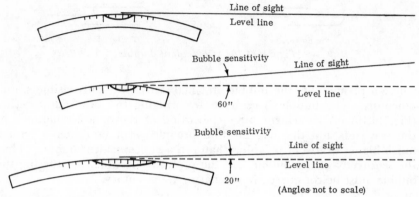

Figure 14.8. The vertical angle through which the line of sight rotates when bubble is off-center one division denotes the sensitivity of the bubble.

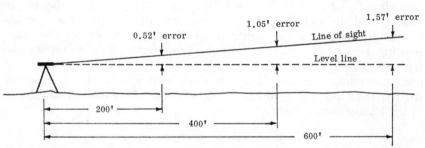

Figure 14.9. Expected error from a level with a 9-minute bubble if the bubble is off-center one division.

TABLE 14.1. Expected Error When the Bubble is a Quarter Division Off-center

Distance— Instrument to Rod (ft)	Farm Level Bubble sensitivity 9 min	Engineer's Level Bubble sensitivity		
		90 sec	1 min	20 sec
200	0.13	0.022	0.015	0.0049
400	0.26	0.044	0.029	0.0097
600	0.39	0.065	0.044	0.0150
1,200	0.78	0.13	0.087	0.0290

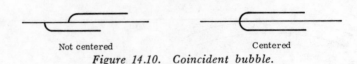

Not centered   Centered

Figure 14.10. Coincident bubble.

where a prism splits the bubble and makes the two ends visible simultaneously. The bubble is centered by bringing the two ends together (Fig. 14.10). A coincident bubble, because of the opposite motions of the two ends and the magnification provided, can be centered with a much higher degree of precision than can a noncoincident bubble. Thus the sensitivity of the bubble and the precision possible in centering the bubble must be considered in estimating the accuracy.

POWER OF THE TELESCOPE

The telescope on farm levels usually has a magnification power of about twelve diameters for the less expensive models to twenty diameters for more expensive models.

Engineer's levels vary in power from twenty to forty diameters. Some engineer's levels are equipped with variable power optics, making possible a variable magnification in the same telescope. Use low magnification with its wide field of view and good illumination for close objects and objects in a dim light. Use high magnification for long-range sighting and precise rod readings in good light. The power of the telescope should be correlated with the sensitivity of the level vial. If a high-power telescope were mounted on a level with a vial of low sensitivity, the tendency would be to take longer shots than the accuracy of the bubble would justify.

QUALITY OF THE OPTICS

The optics of the telescope must be of the quality desired to show the position of the cross hairs on the rod with the desired clarity and precision. Higher quality telescopes have clearer images and greater accuracy, but they are more expensive to manufacture.

WEIGHT, APPEARANCE, AND EASE OF USE

An instrument should be lightweight, easy to attach to the tripod, easy and quick to level, and have an attractive appearance.

COST

The cost should be considered when selecting a level. However, because of the wide range in type and quality, be sure that levels of equal capability are being considered when comparing costs.

SETTING UP

The level head should be taken from its box and mounted securely on the tripod. The tripod leg clamps, which are loosened for storage, should be tightened to provide a firm setup. The tripod should be set up so that the head is approximately level to avoid excessive manipulation of the leveling screws in completing the setup. A quick way to accomplish this is to set two tripod legs firmly in the ground at about the same elevation on the downhill side if the land is sloping. The telescope is turned parallel to these two legs, and the third leg moved one way or the other around the tripod head until the bubble is approximately centered. The telescope is rotated in line with the third leg and this leg moved in or out until the bubble is near the center when the leg is set firmly in the ground.

To complete the leveling process for a level having four leveling screws, the following procedure should be used: Turn the telescope so that it is over two diagonally opposite screws. Turn these screws the same amount but in opposite directions until the bubble is approximately centered. When the screws are turned properly the thumbs move toward or away from one another. The bubble moves in the direction of the movement of the left thumb. Both screws should be firm *but not tight*. If they are too tight or too loose, turn one screw without moving the other until corrected. Rotate the telescope so that it is over the other pair of leveling screws, and bring the bubble to the center. Repeat the process until the bubble remains centered. *Caution:* Tighten the leveling screws only to a firm bearing; excessive tightening may damage the level.

If the level has only three leveling screws, turn the telescope until it is parallel with two of the screws and bring the bubble to the center using either one or both of these screws. Rotate the telescope until it is over the third screw, and level it by using this screw. Repeat this procedure until the bubble remains centered in any position.

CARE

The level is a delicate precision instrument and must be handled carefully. Rough handling may change the relative position of the telescope and bubble tube and cause the level to give incorrect results.

The level head should be carried in its box and placed on the seat of the car or truck or placed in a specially constructed cushioned box when being transported. When moving the level through brush, trees, or other obstructions, it should be carried under the arm with the level head forward. This avoids the possibility of the telescope catching on or striking obstructions. The tripod legs should be spread sufficiently and set firmly into the ground to assure a solid setup and avoid damage that is sure to result if the instrument falls or is blown over. It should be protected from rain or extremely dusty conditions and should not be left set up unattended where it may be overturned or damaged by traffic or livestock.

LEVELING RODS

The leveling rod is used to measure the difference in elevation between the line of sight through the telescope and the point on which the rod is placed. It is usually graduated in feet, tenths, and hundredths of a foot. There are many types of rods available. Two that are widely used are the Philadelphia type (Fig. 14.11) on which a target may or may not be used and the Frisco or California types, which do not permit the use of a target. Each consists of sliding sections permitting the rod to be reduced in overall length for transporting or storing.

When conditions and equipment permit, the rod may be read directly by the instrument man. Figure 14.12 shows the rod as it appears to the instrument man. At this reading the cross hair crosses the rod slightly more than 0.05 foot above the 5-foot mark on the rod. If the accuracy of the survey requires reading the rod to the nearest hundredth of a foot, this reading would be recorded as 5.05 feet. If accuracy to the nearest tenth of a foot is required, it would be recorded as 5.1 feet. Note that the even-numbered hundredths are always at the top edge of the horizontal black bars and the odd-numbered hundredths at the bottom edge.

A target on the rod is useful when staking a line to a specific grade or when reading in a dim light or through foliage.

When holding the rod for a sighting, the rodman faces the level and holds the rod in a vertical position directly in front of himself. It is balanced between the tips of the fingers to assure that it is vertical.

CHECKING AND ADJUSTING

The design of most water management practices is based on surveys. A mistake in the survey may be carried through into the design and construction causing partial or total failure and loss of investment. Such a mistake can come from careless work with a well-adjusted level or from using one that is out of adjustment.

A level is sometimes put out of adjustment by rough handling, but even if it is given good care it should be checked periodically. Any instrument that has been subjected to a serious bump or jar or transported over a long distance should be checked for adjustment before being used.

CROSS HAIRS

The cross hairs are placed in the telescope in such manner that they are always at right angles to each other. The instrument is checked to

Figure 14.11. Philadephia leveling rod. (Courtesy Teledyne Gurley, Troy, N.Y. 12181)

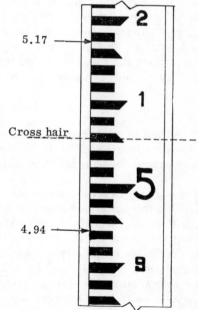

Figure 14.12. Reading a leveling rod.

determine if the horizontal cross hair is truly horizontal. Level the instrument carefully and select a point, perhaps on a post or building, that is directly on the horizontal hair; then rotate the telescope so that this point moves from one side of the magnified field to the other. If the horizontal hair stays on the point through its full length, the cross hair is horizontal. If the hair is on the point at one end and off at the other, it is not horizontal and should be adjusted. The cross hairs are mounted in a cylinder or ring which can be rotated until the horizontal hair is level. Readings made at the exact midpoint of the cross hair are accurate even if it is not level; it is desirable however to adjust the cross hairs so that correct readings may be made anywhere on the horizontal cross hair.

BUBBLE TUBE

When in adjustment the bubble tube is perpendicular to the vertical axis. In checking to determine if it is in adjustment, the instrument is leveled and the bubble accurately centered in the bubble tube. The telescope is then rotated through 180 degrees by turning it end for end. If the bubble remains in the center of the tube at the new position, the tube is perpendicular to the vertical axis. If it has moved from the center in the reversed position, it should be adjusted by bringing it back one-half the distance by adjusting the screws at the end of the tube. Level the instrument and repeat the test, readjusting if necessary, until the bubble remains centered when the telescope is reversed.

LINE OF SIGHT

When in adjusment, the line of sight is parallel to the bubble tube. The procedure to follow in checking the line of sight and making adjustments varies with the type of level.

Turret Farm Level. Remove the screws holding the turret in place and sight on the center of a target or any definite point 150 to 200 feet away. Lift the turret straight up, turn it over, and place it on the plate upside down. Sight toward the target or checking point. If the horizontal cross hair is on the target or point as it was with the turret right side up, the line of sight is parallel to the bubble tube. If the cross hair is not on the point, adjust by bringing the cross hair to a point one-half the distance between. On some instruments the cross hair is shifted by adjusting screws on the cross-hair ring. On others, one end of the telescope is moved up or down. Repeat the whole process as a check.

Dumpy Level. Select two solid points with a difference in elevation of several feet between them, spaced about the same distance apart as the length of sight to be taken with the level, points A and C (Fig. 14.13).

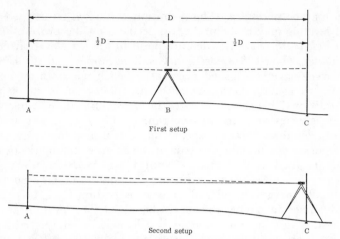

Figure 14.13. Checking the line of sight of a dumpy level. The dotted line represents the line of sight of a level that is out of adjustment. The solid line represents the adjusted line of sight.

If a single check of the level is to be made at this location, points *A* and *C* can be the tops of stakes driven into the ground. If it is to be checked frequently, it is advisable to establish permanent points which will not change in elevation. Set the level at *B* midway between these points and take rod readings on *A* and *C*. Subtract one reading from the other to obtain the difference in elevation between *A* and *C*. The correct difference in elevation between these points is obtained even though the instrument is out of adjustment, for if the lengths of the sights are equal, the same error will be made for each reading if the bubble is exactly centered in each case.

Move the instrument to one of the end points, *C* in Figure 14.13, and set it up so that the eyepiece is less than 1 inch from the rod set on the point. Read the rod by looking backward through the telescope. Even if the instrument is out of adjustment, the error in this short distance will be so small that it can be neglected. The rod reading at *C*, plus or minus the correct difference in elevation between *A* and *C*, will give the correct rod reading at *A*. If the reading at *A* is not correct, the level is out of adjustment and the line of sight must be adjusted until the correct reading is obtained. On some levels the line of sight is adjusted by moving one end of the telescope up or down; on most levels it is adjusted by moving the cross-hair ring up or down.

DIFFERENTIAL LEVELING

The objective of differential leveling is to determine the difference in elevation between points. If the points are located far apart or with an appreciable difference in elevation between them, several setups of the level may be necessary. The definition of several terms used in differential leveling may be helpful in understanding the process.

A *bench mark (BM)* is a definite point of known elevation not subject to change. The U.S. Coast and Geodetic Survey and the U.S. Geological Survey have established permanent bench marks throughout the United States, usually consisting of bronze plates set in stone or concrete. Additional bench marks have been established by other federal and state agencies and private companies. Their elevation above sea level can be obtained from the agency that established the bench mark or usually from an engineer or surveyor in the area. Bench marks of known elevation are used as the starting point for surveys extending over a large area or for those surveys requiring a common reference point.

For many surveys in soil and water conservation work, there is little if any advantage in having the survey based on the elevation above sea level. In such case, an independent bench mark is established for that survey and is given an assumed elevation (usually 100.00 feet). A notch on a tree root, a spike driven into a tree, the high point on a solid stump, the high point on a large boulder, and any other well-defined point that is not subject to change in elevation can be bench marks.

Surveying field notes are the record of work made in a *field book*. Field notes for each job should contain the following information in the upper right-hand corner of the right page: date, description of the weather, names of individuals in the party and their responsibilities, and the instrument used. The description and location of the job are given across the top of the pages (Fig. 14.14).

The form for recording field data varies with the work. Five essential requirements of field notes are accuracy, completeness, legibility, arrangement, and clarity.

Level notes. A standard form for recording level notes for differential leveling is shown in Figure 14.14.

A *plus sight (+S)* is a rod reading taken on a point of known or assumed elevation. The plus sight added to the elevation of the point gives the height of instrument.

The *height of instrument (HI)* is the elevation of the line of sight when the instrument is level.

A *minus sight (−S)* is a rod reading taken on a point whose elevation is to be determined. The minus sight subtracted from the height of instrument gives the elevation of the point.

A *turning point (TP)* is a solid, well-defined point on which the rod is set while the instrument is being moved from one location to another. A minus sight is taken on a turning point from the first level setup to determine its elevation. A plus sight is taken on the turning point from the second setup to determine the new height of instrument.

Elevations are recorded in the fifth column of the field notes.

Procedure: Bench mark 1 (Fig. 14.14) has an elevation of 100.00. Determine the elevation of *BM*-2.

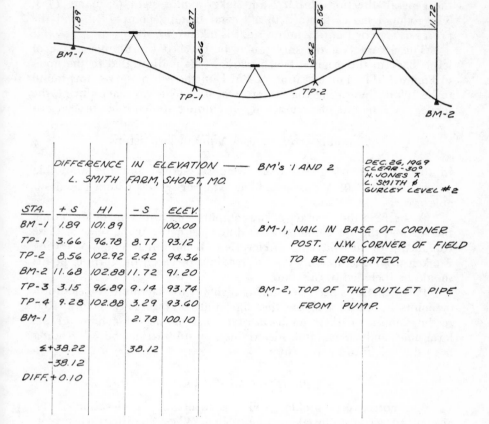

Figure 14.14. Differential leveling, sketch, and level notes.

The rodman holds the rod on *BM*-1 and the instrument man proceeds in the general direction of *BM*-2, the distance depending upon the accuracy of the level, the power of the telescope, and the slope of the land. The level is set up so that a clear view can be obtained of the rod on *BM*-1 and in the general direction of *BM*-2. The rod reading (1.89) on *BM*-1 is recorded in the notes as a +*S* opposite *BM*-1. This added to the elevation of *BM*-1 gives the *HI* (101.89).

The rodman moves in the general direction of *BM*-2 and chooses a turning point—a solid, well-defined point such as a rock, tree root, or the top of a stake driven solidly into the ground, located so that the rod can be clearly seen from the level setup and allowing the rodman an unobstructed view in the general direction of *BM*-2. The rod reading on the turning point (8.77) is recorded in the notes as a −*S* on *TP*-1. This reading subtracted from the *HI* (101.89) gives the elevation of *TP*-1 (93.12).

The instrument man then moves the level to another location in the general direction of *BM*-2 and takes a plus sight (3.66) on *TP*-1. The rodman moves forward to a second turning point *TP*-2, and the process is repeated until a minus sight is taken on *BM*-2.

The accuracy of the survey may be checked by running a line of levels back to the original bench mark. This is illustrated in the notes in Figure 14.14. The elevation of *BM*-1 on the return survey was found to be 100.10, indicating that an error of 0.10 foot was made during the survey. A standard of accuracy for beginning students in surveying is

$$\text{Allowable error} = 0.05 \sqrt{\text{number of setups}}$$

In the set of notes given, there were six setups and the allowable error would be $0.05 \sqrt{6}$ or 0.12 foot. The survey was within the allowable error.

A check on the accuracy of the computations can be made by adding the $+S$ and the $-S$ columns. The difference between the sums of these columns gives the difference in elevation. If more than one $-S$ reading is taken from a setup, only the $-S$ readings taken on the turning points should be included in the sum.

It is sometimes necessary during a survey to determine the elevation of points in addition to the turning points or the bench marks. Rod readings taken on these points are recorded in the $-S$ column of the level notes and the elevation determined by subtracting the $-S$ reading from the height of instrument.

PROFILE LEVELING

The objective of profile leveling is to obtain the elevation of the ground surface at points along a given line. These elevations when plotted give a graphical representation—a profile of a vertical section of the ground surface along this line. A profile is useful in determining the grade and depth of cut required in constructing soil and water conservation practices such as drainage ditches, diversions, irrigation canals, and tile drains.

Procedure: Staking the line. The general location of the improvement is selected. Stakes are driven along the proposed center line at 100-foot intervals and at intermediate points if there is a change in the ground slope that will not be shown by the 100-foot stakes. Each 100-foot point is called a station. The stake at the beginning of the survey is marked $0 + 00$. The station stakes are marked $1 + 00$, $2 + 00$, etc.; an intermediate stake marked $2 + 85$ indicates that this stake is 85 feet beyond $2 + 00$, or 285 feet from the beginning of the survey.

Obtaining elevations of the ground surface. First, a bench mark, or reference point is established. The level is set up near but not on the survey line, and a plus sight is taken on the bench mark, correct to the

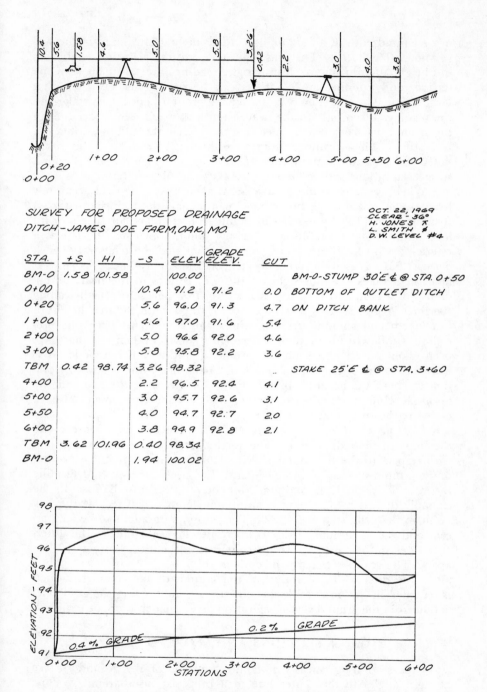

Figure 14.15. Profile leveling sketch, survey notes, and profile.

nearest hundredth of a foot, and recorded in the $+S$ column of the survey notes (Fig. 14.15). The plus sight added to the elevation establishes the height of instrument. Minus sight readings are taken on as many stations and intermediate points as can be read accurately from that level setup. The rod should be set on the ground beside each stake and the readings taken only to the nearest tenth of a foot, since variation in the ground surface is usually such that more accurate readings would not be justified. These readings are recorded in the $-S$ column of the survey notes and subtracted from the HI to obtain the elevations. Ground surface elevations should be computed to the nearest tenth of a foot.

When as many readings have been taken as can be read accurately from the first setup, the instrument man signals the rodman to take a turning point (TP) by moving his hand in a circular motion above his head. The rodman establishes a turning point, which may be left as a reference point for future use; if so, it is designated as a temporary bench mark, TBM. It is usually not located on the survey line. The instrument man takes a minus sight on the TP, correct to the nearest hundredth of a foot, and then moves the level to a new setup further along the line. A plus sight is taken on the TP to establish the new height of instrument and another series of minus sights is taken. The same procedure is followed for the remainder of the survey. When readings have been taken on all points on the profile, a line of differential levels is run back to the original bench mark or to another point of known elevation to check the accuracy of the survey. A sketch of the site and the profile leveling notes taken in making a survey for a proposed drainage ditch are shown in Figure 14.15.

Plotting the profile. The elevations of the ground surface along the line are plotted to scale. The vertical scale is usually exaggerated with respect to the horizontal scale, so that variations in the elevation of the ground surface are more evident. The profile is used to determine the most desirable depth and grade for the proposed installation. In the example the ditch must have a minimum depth of 2.0 feet at the low point in the field, station $5 + 50$. The minimum grade is to be 0.2 percent and the maximum grade, 0.4. percent. Percent grade is the difference in elevation (in ft) between two points in the ditch bottom 100 feet apart. The grades selected will result in the least depth of cut at points along the ditch. The elevation of the bottom of the completed ditch is recorded in the notes as "grade elevation." Subtracting the grade elevation from the ground surface elevation gives the depth of cut required.

CROSS SECTIONS AND SLOPE STAKES

Cross-sectioning consists of taking a profile of the ground surface at right angles to the center line of a proposed installation. A slope stake is set at the point where a cutslope, in the case of an exacavation, or a fill slope, in the case of an embankment, intersects the original ground.

EXCAVATIONS

Most surveys for cross-sectioning and setting slope stakes are made after profile levels have been run and the grade elevation and depth of cut computed. The survey for cross-sectioning and setting slope stakes at two stations on the drainage ditch of Figure 14.15 is shown in Figure 14.16. A rod reading (2.66) is taken on the bench mark, *BM*-0 and the *HI* computed (102.66). The grade elevation (91.3 at station $0 + 20$) is obtained from the profile notes (Fig. 14.15). The difference in elevation between the grade elevation and the *HI* (11.4) is recorded in the column headed "Grade Rod." The grade rod is the rod reading that would be obtained if it were taken in the bottom of the finished ditch. Any rod reading taken on the ground surface when subtracted from the grade rod gives the difference in elevation between the ground surface and the bottom of the ditch. This is referred to as the *depth of cut.*

CROSS SECTIONS ON DRAINAGE DITCH						DITCH 6' BOTTOM 2:1 SIDE SLOPES		OCT. 23, 1969 CLOUDY-55° H. JONES ? L. SMITH ? D. W. CLEVEC #4		
JAMES DOE FARM, OAK, MO.										
STA.	+S	HI	-S	ELEV.	GRADE ELEV.	GRADE ROD		LEFT	CENTER	RIGHT
BM-0	2.66	102.66		100.00						
0+20			6.7	96.0	91.3	11.4	CUT ROD	6.0 / 5.4 / 15.0	4.7 / 6.7 / 0	4.2 / 7.2 / 11.4
1+00			5.7	97.0	91.6	11.1	CUT ROD	6.2 / 4.9 / 15.4	5.4 / 5.7 / 0	5.1 / 6.0 / 13.2

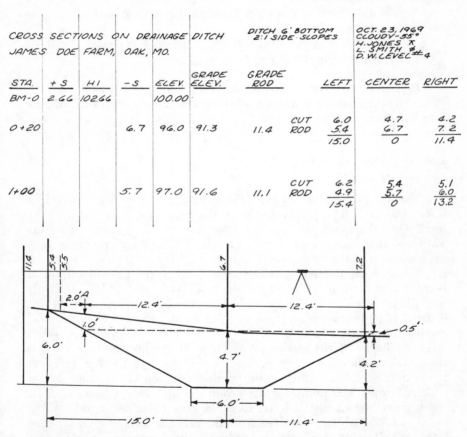

Figure 14.16. Survey for cross-sectioning and setting slope stakes on drainage ditch illustrated in Figure 14.15. The sketch shows the cross section at station $0 + 20.$

The cross-section notes are recorded on the right-hand page of the field book as a series of fractions. The distance that the rod reading was taken to the right or left of the center line is recorded as the denominator. Either the rod reading, the depth of cut, or both are recorded as the numerator of the fraction. In Figure 14.16 both the rod reading and the depth of cut are recorded.

Slope stakes should be set at the points where the cutslopes intersect the original ground. They are set by a trial-and-error process. The distance that a slope stake is set from the center line can be computed by

$$d = \frac{w}{2} + hz$$

where $w =$ bottom width of ditch

$h =$ difference in elevation between the ground surface and the bottom of ditch

$z =$ the side slope ratio (ratio of horizontal to vertical); for a two-to-one slope, $z = 2$

Procedure: See Figure 14.16 and the following:

First, the distance d using h at the center line is computed for station $0 + 20$.

$$d = \frac{6}{2} + 4.7 \times 2 = 12.4 \text{ feet}$$

The cutslope would intersect the ground 12.4 feet from the center line if the land slope perpendicular to the center line of the ditch were level. On the left side of the center line however the ground slopes upward and if the slope stake were set at A, 12.4 feet from the center line, the side slope of the ditch would be steeper than two to one.

Estimate the difference in elevation between the ground surface at point A and the center line of the ditch (1.0 ft in Fig. 14.16). Multiply this difference by the side slope ratio ($2 \times 1 = 2$) and move the rod this distance (2 ft) beyond point A, or 14.4 feet from the center line, and take a rod reading. The reading (5.5) subtracted from the grade rod (11.4) gives the depth of cut at this point (5.9). With this depth of cut the slope stake should be set at a distance $d = 6/2 + 5.9 \times 2 = 14.8$ feet from the center line, which indicates that the slope stake must be set slightly beyond this point, for the ground is still sloping upward. The rod is moved to a point 15.0 feet from the center line and a rod reading of 5.4 is taken, which subtracted from the grade rod (11.4) gives a cut of 6 feet at this point. With this depth of cut the slope stake should be set at a distance $d = 6/2 + 6.0 \times 2 = 15.0$. The distance the rod reading was taken from the center line was also 15 feet, so the slope stake is driven at this point, and the depth of cut and distance from the center line is recorded in the cross-section notes.

The process is repeated on the right side of the ditch. At a point 12.4 feet from the center line, it is estimated that the ground surface is 0.5 foot lower than the ground surface at the center line, so the ditch slope will intersect the ground surface closer to the center line. The rodman moves back twice the estimated difference ($2 \times 0.5 = 1.0$) to a distance 11.4 feet from the center line and takes a rod reading, 7.2 feet, which, subtracted from the grade rod (11.4), gives a depth of cut of 4.2 feet. With this depth of cut the slope stake should be $d = 6/2 + 4.2 \times 2 = 11.4$ feet from the center line—the same distance at which the rod reading was taken—so the slope stake is driven at this point and noted in the cross-section notes.

EMBANKMENTS

After the center line of an embankment has been staked out and the elevation of the top of the embankment determined, cross sections are taken to determine the volume of earth needed, and slope stakes are set to outline the base.

In Figure 14.17 the level has been set up so that the line of sight is 6.7 feet below the top of a proposed embankment. The rod reading on the top would be -6.7 feet. A rod reading of 7.1 feet was taken at the center line of the embankment at station $0 + 38$. The height of fill above this point is $7.1 - (-6.7) = 13.8$ feet. Because of the variation in the ground slope perpendicular to the center line, rod readings were also taken at distances of 6 feet and 20 feet upstream and 15 feet downstream from the center line. These are recorded in the notes, and the height of fill above these points is computed.

The distance from the center line that a slope stake for an embankment should be set can also be computed by the formula

$$d = \frac{w}{2} + hz$$

where $w =$ top width
$h =$ height of the top of the fill above the point
$z =$ side slope ratio

In the cross section in Figure 14.17, the front slope would intersect the original ground at a point $d = 12/2 + 13.8 \times 3 = 47.4$ feet from the center line if the ground were level. At a point 47.4 feet from the center line it was estimated that the ground surface was 4 feet higher, so the rodman moved toward the center line a distance equal to 3 times 4, or 12 feet. A rod reading (2.0) was taken at the point 35.4 feet from the

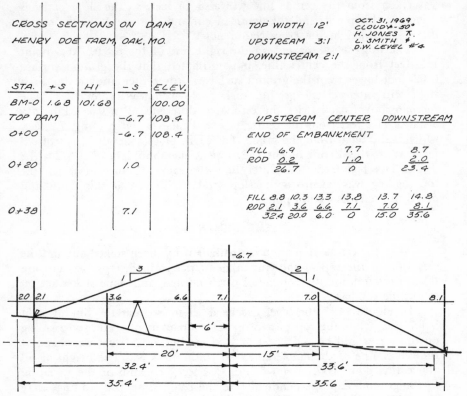

STA.	+ S	HI	− S	ELEV.
BM-0	1.68	101.68		100.00
TOP DAM			−6.7	108.4
0+00			−6.7	108.4
0+20			1.0	
0+38			7.1	

CROSS SECTIONS ON DAM
HENRY DOE FARM, OAK, MO.

TOP WIDTH 12'
UPSTREAM 3:1
DOWNSTREAM 2:1

OCT. 31, 1969
CLOUDY - 50°
H. JONES ⚲.
L. SMITH ⚲
D.W. LEVEL #4

UPSTREAM   CENTER   DOWNSTREAM
END OF EMBANKMENT

FILL 6.9   7.7   8.7
ROD 0.2   1.0   2.0
     26.7   0   23.4

FILL 8.8 10.3 13.3 13.8   13.7 14.8
ROD 2.1 3.6 6.6 7.1   7.0   8.1
    32.4 20.0 6.0 0   15.0 35.6

Figure 14.17. Survey for cross-sectioning and setting slope stakes for an embankment. The sketch shows the cross section of the embankment at station 0 + 38.

center line. The height of fill above this point was 8.7 feet, and the slope stake should be set a distance $6 + 8.7 \times 3 = 32.1$ feet from the center line. The rodman moved to a point 32.1 feet from the center line and obtained a rod reading of 2.1 feet, giving a height of fill of 8.8 feet. With a height of fill of 8.8 feet, the slope stake should be set at $6 + 8.8 \times 3 = 32.4$ feet. If there is no significant difference in elevation between this point and the point of the previous rod reading, another reading is not needed, and the slope stake is set at 32.4 feet.

The backslope would intersect the ground at a point $6 + 13.8 \times 2 = 33.6$ feet from the center line if the ground were level. The ground surface at this point is estimated to be 1 foot lower than at the center line, so a rod reading is taken 2 feet farther from the center line at 35.6 feet. The reading at this point is 8.1. The height of fill above this point is 14.8 feet and the slope stake should be a distance of $6 + 14.8 \times 2 = 35.6$ feet from the center line. This is the point at which the rod reading was taken, so a slope stake is driven at this point.

PROBLEMS

14.1. In adjusting a dumpy level the rod reading obtained on point A from the center setup (Fig. 14.13) was 4.24 feet and on point C it was 5.78 feet.

a. What is the difference in elevation between A and C?

b. The level was moved to point C and the rod reading on point C was 4.68 feet and on point A, 3.54 feet. Was the level in adjustment? If not, how is it adjusted?

c. When the line of sight is corrected, what will be the rod reading on C? Explain.

d. If the level had been in adjustment, what would have been the rod readings on point A and point C from the center setup?

TABLE P14.1. Survey Notes for Problem 14.2

Station	+S	HI	−S	Elevation
BM-6	10.86			100.00
TP-1	0.70		1.62	
TP-2	9.63		11.61	
TP-3	4.50		1.25	
BM-7	5.05		5.06	
TP-4	0.34		4.66	
TP-5	11.22		10.24	
TP-6	2.64		0.46	
BM-6			9.92	

14.2. Record the survey notes given in Table P14.1 in a field book and compute the elevation of all points. Is the survey within the allowable error for beginners? Describe, in general terms, the topography of the area over which it was made.

14.3. The rod readings taken during a level survey are indicated in Figure P14.1.

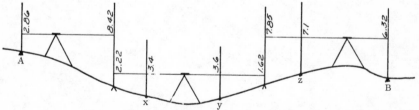

Figure P14.1. Sketch for use in Problem 14.3.

a. Record the rod readings in a field book using the form suggested for level notes. Assume the elevation of point A to be 100.00 feet. Determine the elevation of points x, y, z, and B.

b. Could the difference in elevation between points A and B have been determined by adding the $+S$ and $−S$ columns and taking the difference? Explain.

14.4. The survey notes in Table P14.2 were taken during the survey for a proposed drainage ditch. Record these notes in a field book, compute the eleva-

TABLE P14.2. Survey Notes for Problem 14.4

Station	+S	HI	—S	Elevation	Grade Elevation	Cut
BM	2.62			100.00		
0 + 00			12.6			
0 + 30			9.5			
1 + 00			8.5			
2 + 00			9.0			
TP-1	8.02		7.24			
3 + 00			9.3			
4 + 00			10.0			
TP-2	7.21		8.82			
5 + 00			7.3			
6 + 00			5.5			
7 + 00			6.3			

tion of all points, and plot the profile of the ground surface. The elevation of the ditch bottom at the outlet, 0 + 00, is to be 90 feet. The maximum grade of the ditch is to be 0.5 percent and the minimum grade, 0.2 percent. The ditch is to be not less than 2 feet deep at any point. Plot the profile of the ditch bottom that will result in the least amount of work in digging the ditch. Indicate on the profile the grade used on each section of the ditch. Compute grade elevation and depth of cut at each station.

14.5. Compute the number of cubic yards of earth to be removed from the drainage ditch in Figure 14.16 between stations 0 + 20 and 1 + 00.

14.6. Compute the number of cubic yards of earth required in the embankment in Figure 14.17 between stations 0 + 20 and 0 + 38.

15 ～ Land Surveys, Topographic Maps, and Aerial Photographs

Any person working on plans involving property lines should have a knowledge of the United States system of public land surveys.

IN THE OLDER portions of the United States, streams, ridge lines, and other natural features were accepted as property lines. As the country developed and tracts of land were subdivided it became the practice to determine the lengths and direction of property boundaries with a surveyor's chain and compass, and to fix the location of property corners by artificial monuments. When the direction and lengths of the sides are given, the property is said to be described by "metes and bounds."

In 1785 the Continental Congress passed an ordinance that provided for the subdivision of public lands by the rectangular system which in general principle was the system followed after that date.

The land is divided into townships and sections which are located with respect to principal axes passing through an *initial point,* which is the point from which the surveys for that particular area were started. The north-south axis is a true meridian called the *principal meridian,* and the east-west axis is a true parallel of latitude called the *base line.* (See Fig. 15.1.)

Standard parallels of latitude are located at intervals of 24 miles north and south from the base line. Guide meridians are located by measuring 24 miles east or west of the guide meridian along the base line and along each of the standard parallels and then running a true meridian from that point. Because of the convergence, the north side of each 24-mile quadrangle is less than 24 miles in length.

The 24-mile quadrangles are divided into townships by *range* and *township lines.* At intervals of 6 miles measured along each standard parallel, a range line is run as a true meridian to the next standard

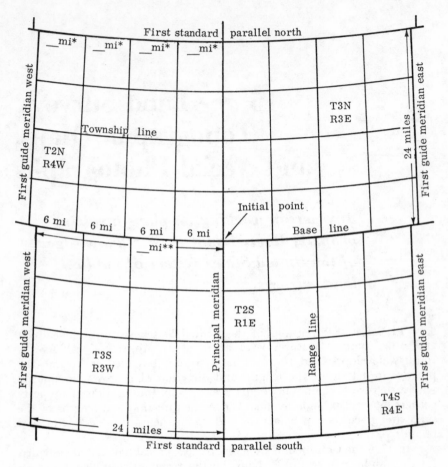

* 6 miles less convergence

** 24 miles less convergence

Figure 15.1. Rectangular system of land surveys.

parallel north. At intervals of 6 miles measured along the guide meridian, a township line is run parallel to the standard parallel. A township is established which is 6 miles on each side, except for the north
side which is less than 6 miles due to convergence of the range lines.
Townships are designated by the number of townships and ranges, their
direction from the initial point, and number or name of the principal
meridian. For example, T3N, R6W, 5th principal meridian, designates
a township in the third tier of townships north of the base line and in
the sixth range west of the fifth principal meridian.

The townships are divided into sections as indicated in Figure 15.2.
All sections will be 1 mile square, if no errors are made in the surveys,

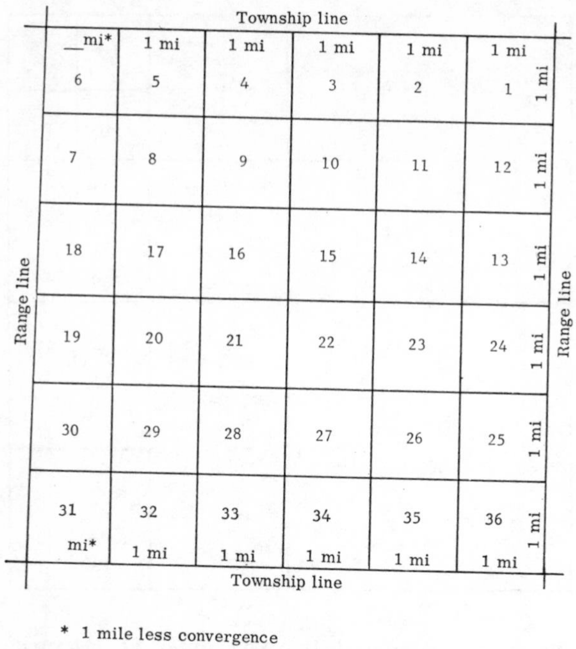

Township line					
__mi*	1 mi	1 mi	1 mi	1 mi	1 mi
6	5	4	3	2	1
7	8	9	10	11	12
18	17	16	15	14	13
19	20	21	22	23	24
30	29	28	27	26	25
31	32	33	34	35	36
mi*	1 mi	1 mi	1 mi	1 mi	1 mi
Township line					

Figure 15.2. Subdivision of a township into sections.

Range line (left), Range line (right), 1 mi markings on right column.

* 1 mile less convergence

except those along the west boundary which will be less than 1 mile in width because of convergence of the range lines.

The sections may be further divided into smaller tracts as indicated in Figure 15.3. The legal description of these subdivisions always begins with the smallest unit. For example, if section 16 in Figure 15.3 were located in T3N, R6W, the legal description of the 10 acres designated by X would be SE ¼, NE ¼, NE ¼, Sec 16, T3N, R6W.

Additional information on public land surveys is available in most textbooks on surveying, such as those listed as references.

TOPOGRAPHIC MAPS

The topographic map is a small-scale representation of the topography, or relief, of an area. It may also show natural features such as trees and streams, and artificial features such as lakes, buildings, and roads.

Topographic maps are used in many ways. They are useful in the design of any project which requires consideration of the shape of the earth's surface. In soil and water conservation work, it would be an advantage to have a topographic map of a farm on which a water management system is to be planned. If the map had the proper scale and contour interval, rearrangement of fields and location of terraces, waterways, and diversions could be plotted on the map and the most desirable arrangement developed.

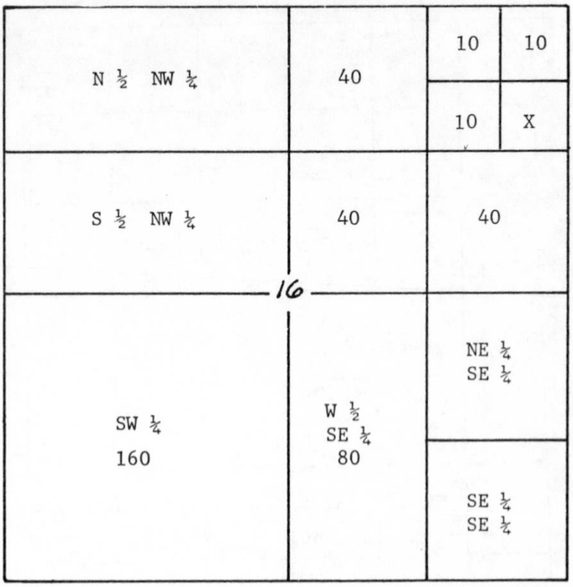

Figure 15.3. Subdivision of a section.

N ½ NW ¼	40	10	10
		10	X
S ½ NW ¼	40	40	
	— 16 —		
SW ¼ 160	W ½ SE ¼ 80	NE ¼ SE ¼	
		SE ¼ SE ¼	

Most topographic maps are prepared by governmental organizations. The U.S. Geological Survey has published topographic maps of most of the United States. Other agencies such as the U.S. Army Engineers and U.S. Department of Agriculture have maps of specific areas. The central source of information on maps prepared by federal agencies is the Map Information Office, U.S. Geological Survey, Washington, D.C. In addition, many maps are available from state, county, and city agencies. These maps may not be made to the best scale or contour interval for planning soil and water management practices in detail; however, they may be useful in planning the overall system.

A topographic map that has been prepared for study is shown in Figure 15.4.

Contours and contour lines. A contour is an imaginary line that passes through points of equal elevation on the ground surface. A contour line is a line on a map passing through points of equal elevation. The elevation of the contour is given by a number which appears on selected contour lines. The difference in elevation between successive contour lines is called the contour interval. The contour interval selected for a map depends upon the degree of accuracy desired, the steepness of the slope, the scale of the map, and the cost.

Contour lines are close together on steep slopes and farther apart on flat slopes. They point upstream in crossing a valley or stream. The steepest land slope is always at right angles to a contour line. On uniform slopes, the contour lines are spaced uniformly. All contour lines must close either within or without the boundaries of a map. A closed

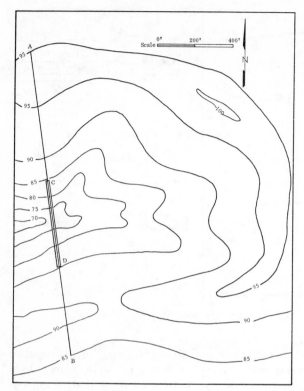

Figure 15.4. Topographic map for study.

contour line, which has no other contours within, represents either a summit or a depression.

Map scale. The scale to which a map is drawn is determined by the use to be made of the map, the size of the area, and the detail desired.

Map symbols. Objects are represented on the map by symbols. Some of those commonly used in mapping for soil and water conservation work are shown in Figure 15.5.

MAKING TOPOGRAPHIC MAPS

If the topographic map that is available for an area does not have the desired contour interval, scale, or detail needed, it may be necessary to make a map by one of the following methods:

Grid. This method of mapping is adapted to areas that are to be land graded for irrigation or drainage. The area to be mapped is laid out in a grid pattern, either square or rectangular. The size of the grid will depend upon the size of the area, the steepness and irregularity of the slope, and the accuracy desired. Stakes may be set at all grid points, or two lines of stakes may be set at right angles to each other

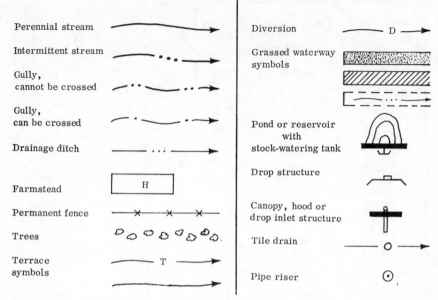

Figure 15.5.   Commonly used map symbols.

and additional grid corners located by alignment on the appropriate pair of stakes. The staked corners are indicated by O in Figure 15.6. The grid corners may be identified by the numbering and lettering system indicated. The ground elevations at each grid corner are obtained with a level and recorded, and contour lines drawn at the desired contour interval.

Outline map. An outline map of the area showing boundaries, fences, ditches, and other features is needed. If such a map is not available, one may be reproduced from an aerial photograph with sufficient accuracy for most purposes. A level or transit is used to obtain the elevations of points in the field that can be located on the map. Elevations at additional points where changes in slope or other variations in topography occur are usually needed. These points can be located by appropriate measurements from other points already established on the map. An outline map on which the elevations have been recorded is shown in Figure 15.7.

Transit-stadia. Points in the field are located by measuring the angle to the point by the transit and the distance to the point by stadia. Elevations are obtained by calculating the vertical distances from the vertical angles and stadia distances.

Level-stadia. This method requires a level equipped with a horizontal circle. Points in the field are located by measuring the angle to the point by the use of the horizontal circle and the distance to the

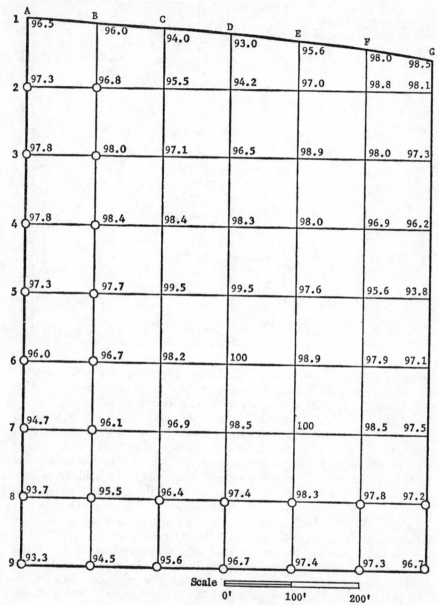

Figure 15.6. Grid layout for a topographic map. Stakes have been set at O; other points are located by alignment.

Figure 15.7. Outline map prepared from an aerial photograph.

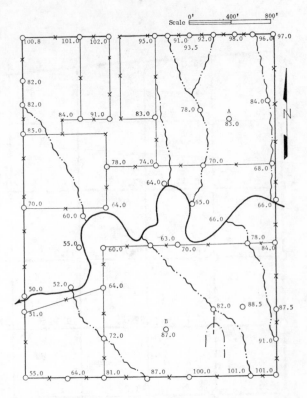

point by stadia. This method is adapted to taking the measurements necessary to prepare a map of terraces as they are being staked out. The map can then be used to plan an improved terrace layout.

Plane table. A plane table consists of a board that is attached to a tripod in such a way that it can be leveled and rotated on a vertical axis. It is used as a drafting table for plotting map details and contour lines in the field. Points to be mapped are sighted through an alidade and the direction to the point noted on the map. An alidade is constructed so that a straightedge which rests on the board is parallel to the line of sight, which is established by peep sights or a telescope. A telescopic alidade also has stadia hairs and a vertical scale so that the distance and elevation of the point can be determined.

Photogrammetry. Photogrammetry, the art and science of obtaining horizontal and vertical measurements from photographs, is rapidly replacing ground surveys as a method of obtaining topographic maps. The precision and accuracy obtained by photogrammetric methods depends on the altitude at which the photographs are taken, the quality of the equipment used, and the skill of the personnel; it is often superior to that obtained by ground survey methods. Many government agencies use photogrammetric methods in preparing topographic maps. Various private companies specialize in aerial mapping and photogrammetry, and their services are being used in planning soil and water conservation systems.

Additional information on the details of topographic mapping by any of the above methods can be found in surveying or photogrammetry textbooks or manuals. See references at the end of this chapter.

AERIAL PHOTOGRAPHS

Aerial photographs such as the one shown in Figure 15.8 can be quite useful in planning soil and water conservation systems. An outline map of an area can be reproduced with sufficient accuracy for most purposes from an aerial photograph. The photographs can be used in determining the drainage area above structures, ponds, and reservoirs and in determining the type of land use and vegetation. Useful information

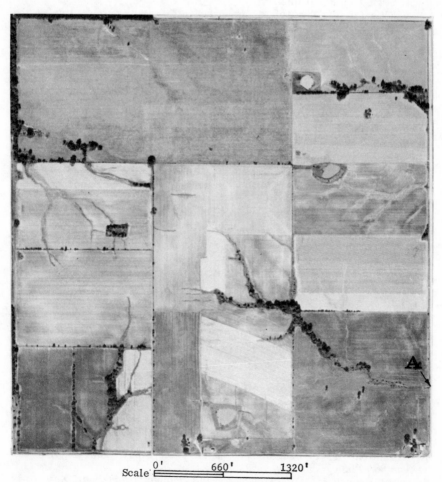

Scale 0'   660'   1320'

Figure 15.8. Aerial photograph.

on the use and interpretation of aerial photographs is given in references listed.

Most of the United States has been photographed for various federal agencies. Information on photographs available may be obtained from the Map Information Office, U.S. Geological Survey, Washington, D.C.

Most of the crop and rangeland in the United States has been photographed for the Agricultural Stabilization and Conservation Service, U.S. Department of Agriculture. Information on how to procure these maps can be obtained from the local ASCS office.

PROBLEMS

15.1. Refer to Figure 15.4. A dam constructed from C to D backs water up to an elevation of 82 feet.
    a. Plot a profile of line A-B.
    b. What is the ratio of the watershed area to the surface area of the reservoir?
    c. What is the average slope of the land draining into the reservoir?
    d. Describe the topography represented on the map.

15.2. Use transparent paper as an overlay and draw in contours on a 1-foot interval on the map in Figure 15.6.

15.3. Refer to the map in Figure 15.7.
    a. Using transparent paper as an overlay, draw in contours on a 10-foot interval on the map.
    b. What measurements would have been taken in the field in order to locate points A and B on the map?

15.4. A structure is to be placed across the gully at point A in Figure 15.8. Use transparent paper as an overlay and trace out the watershed boundary of the area draining into this structure. How many acres are in this drainage area?

REFERENCES

SURVEYS

1. Breed, C. B., and Hosmer, G. L. *Elementary surveying.* Principles and practice of surveying, vol. 1. 10th ed. New York: Wiley, 1966.
2. Brinker, R. C. *Elementary surveying,* 5th ed. Scranton, Pa.: International Textbook, 1969.
3. Davis, R. E., Foote, F. S., and Kelly, J. W. *Surveying theory and practice,* 5th ed. New York: McGraw-Hill, 1966.
4. Rayner, W. H., and Schmidt, M. O. *Fundamentals of surveying.* Cincinnati: Van Nostrand, 1969.

AERIAL PHOTOGRAPHS

5. American Society of Photogrammetry. *Manual of photographic interpretation.* Falls Church, Va.: Am. Soc. Photogrammetry, 1960.
6. Avery, Eugene T. *Interpretation of aerial photographs.* Minneapolis: Burgess, 1968.
7. Smith, John T. *Manual of color aerial photography.* Falls Church, Va.: Am. Soc. Photogrammetry, 1968.
8. Spurr, Stephen H. *Photogrammetry and photo-interpretation.* New York: Ronald Press, 1960.

INDEX